SPACE DEBRIS

Clearing the Final Frontier

JASON P. CRUTCHFIELD

First Edition · 2026

Copyright

First Edition: 2026.

Printed in the United States of America.

To those who believe in the infinite potential of space exploration
and in the responsibility that comes with it.

Preface

High hopes and dreams of exploring beyond our terrestrial bounds marked the dawn of the space age. Since Sputnik 1's launch in 1957, humanity's reach into the cosmos has grown exponentially. Satellites now power our global communications, guide our navigation systems, and provide critical data for scientific research, weather forecasting, defense, and a hundred industries that did not exist when the first artificial object reached orbit. The transformation has been so complete that most users of these services no longer notice the orbital infrastructure they depend on. They notice only when something fails.

This unprecedented growth in space activity has brought a less glamorous byproduct: space debris. Fragments of old satellites, discarded rocket stages, debris from past collisions and explosions, and even tiny flecks of paint hurtling through space at orbital velocities now pose a significant and growing threat to modern space infrastructure. The orbital environment that the early pioneers treated as effectively infinite is, on closer inspection, finite and increasingly congested. Without proper intervention and sustained international collaboration, we risk making the most useful regions of near-Earth space inhospitable to the missions humanity now depends on.

This book explores the origins, risks, and future of space debris in depth. It offers a comprehensive analysis of the current state of the orbital environment, the historical events that shaped it, the technologies and policies that aim to manage it, and the ethical and economic questions that will shape its future. It is intended to serve readers across the full range of audiences with a stake in this question: students entering aerospace engineering, policymakers shaping national and international space governance, industry professionals working in operations, insurance, or law, scientists studying the environment itself, and the broader public whose lives depend on the satellite services enabled by the orbital environment.

The book is designed to be read straight through as a coherent argument and to be useful chapter by chapter as a reference. Each chapter is self-contained yet builds on what precedes it. Technical material is presented at a depth sufficient to support the argument but does not assume specialized training; readers seeking deeper technical treatment will find the references useful for further exploration. Policy and legal material are presented to the same standard: enough detail to understand the substance, supplemented by references to primary instruments and authoritative analyses for those who need them.

Whether you are a student, policymaker, engineer, or simply a space enthusiast, this book aims to educate and inspire. The final frontier may be vast, but it is not infinite. What we do now will shape the future of space exploration for generations.

Acknowledgments

Writing a book about space debris and its far-reaching implications would not have been possible without the support of many individuals and organizations. I want to thank the countless researchers, engineers, and policymakers whose work and dedication to solving the space debris crisis have provided invaluable insights and data. Their published work, often technical and difficult for a general audience to access, forms the foundation of this book. When I have summarized or interpreted their findings, I have tried to do so faithfully, and when I have differed from their conclusions, I have tried to do so transparently.

I am particularly indebted to the staff of the European Space Agency's Space Debris Office, whose annual Space Environment Report is the most authoritative public document on the state of the orbital environment, and to the staff of NASA's Orbital Debris Program Office, whose ongoing publications and quarterly news updates have been essential references throughout. The Inter-Agency Space Debris Coordination Committee, the United Nations Office for Outer Space Affairs, and the International Telecommunication Union have developed the international frameworks that coordinate the global debris response. The technical and legal literature these institutions have produced is the closest the field has to common ground.

I extend my heartfelt gratitude to my family and friends, who supported me throughout the writing process, and to the mentors who instilled in me a passion for space exploration. Special thanks to my wife, Diana, whose feedback and encouragement made this book a reality and whose patience during the long nights of writing exceeded any reasonable expectation.

I also owe a debt of gratitude to the artificial intelligence tools that supported this project from start to finish. What began as a sprawling collection of PowerPoint slides, spreadsheet graphs, and rough tables was transformed, with their help, into figures and visuals that are far more pleasant to look at and much easier to understand. Beyond the visuals, these tools assisted me tremendously throughout the editing process, helping me catch inconsistencies, tighten the prose, and shape a long, technical manuscript into something coherent and readable. The book is better for that assistance, though any errors that remain are entirely my own.

Finally, I am indebted to the readers and to the future leaders in space policy and technology for whom this book is ultimately written. Your commitment to responsible space exploration will shape our shared future beyond Earth.

Note on This Edition

This edition reflects comprehensive editorial work undertaken in 2026 to update the book to align with current authoritative sources, restructure its organization for greater clarity, and expand its coverage to address topics not adequately covered in the first edition. The principal changes include comprehensive fact-checking against the European Space Agency's 2025 Space Environment Report; updated treatment of major debris-generating events in recent years; expanded coverage of reentry hazards and the atmospheric effects of reentered spacecraft; new chapters addressing astronomical interference and the cislunar environment; deeper treatment of national regulatory frameworks beyond the United States; and integration of recent developments in active debris removal, sustainability ratings, and space traffic management.

The chapter structure has been revised into a sixteen-chapter sequence that directly follows the field's substantive logic: from the basic physics through the historical record, the legal and regulatory frameworks, the operational and technological responses, and the longer-term questions of sustainability and equity. Each chapter has been rewritten with attention to consistent voice, accurate detail, and current source material. The references have been consolidated and formatted in APA style. A glossary, complete index, and full back matter accompany the revised text.

When data change, the text reflects the most current available figures, with explicit attribution to the source and date. When subsequent developments supersede the original argument, the text notes both the original position and the reasons for the update. The aim throughout has been to produce a book that serves as a current and complete reference on the state of the orbital debris problem, the responses to it, and the choices that will shape its trajectory in the decades to come.

Contents

FRONT MATTER

CHAPTERS

BACK MATTER

Chapter 1. The Space Debris Crisis: An Introduction

The Hidden Cost of the Space Age

On October 4, 1957, the Soviet Union launched a polished aluminum sphere about the size of a beach ball into low Earth orbit. Sputnik 1 weighed 83.6 kilograms (184 pounds) and carried four whip antennas, two radio transmitters, and a battery that powered three weeks of beeping. It was the first artificial object to orbit Earth, marking the beginning of the space age. A few observers noted at the time that Sputnik 1 was not the only object to reach orbit on that flight. The R-7 rocket's core stage, which had carried the satellite into space, also entered orbit and remained there for two months before reentering. Sputnik 1 itself decayed and burned up on January 4, 1958. The rocket stage was, in a real sense, the first piece of space debris.

Almost seventy years later, the orbital environment that Sputnik inaugurated is no longer a vast, empty volume occasionally crossed by a satellite. According to the European Space Agency's 2025 Space Environment Report, roughly 40,000 objects larger than ten centimeters are currently tracked in Earth orbit by global space surveillance networks. Approximately 11,000 of these are active, functioning satellites. The remainder, including defunct spacecraft, spent rocket bodies, mission-related debris, and fragments from past collisions and explosions, constitutes debris. When statistical models extend the count below the tracking threshold, the picture becomes more dramatic: an estimated 1.2 million objects between 1 and 10 centimeters (0.39 inches to 3.94 inches) in size, and roughly 140 million between 1 millimeter and 1 centimeter

(0.039 inches to 0.39 inches). Each of these, traveling at orbital velocity, is a potential weapon.

The numbers describe a problem that is no longer hypothetical. The space debris environment poses a measurable, growing risk to every satellite in operation, to the International Space Station and its crew, and to the long-term usability of the orbital regions on which modern civilization has come to depend. Yet the problem has developed gradually enough, and the affected systems are remote enough from daily experience, that public awareness has lagged behind the technical reality by decades. This book is about how that environment came to be, what is being done about it, and what must be done to keep low Earth orbit accessible for the missions of the coming decades.

The following chapter outlines the problem in several steps. It begins by defining debris and how it is categorized, by both physical size and origin. It addresses the physics that make orbital debris uniquely dangerous: the high velocities of orbital objects and the consequences of hypervelocity impacts. It surveys the principal sources of debris, including mission-related releases, fragmentation events, and degradation products over time, and identifies the categories of greatest concern. It examines what is at stake for modern civilization in the continued usability of the orbital environment. It introduces the concept of space sustainability and the international frameworks that have emerged to address it. And it provides a map of the rest of the book, identifying which questions are addressed in which chapters.

Defining Space Debris

Space debris, also called orbital debris or simply debris, refers to any artificial object in Earth orbit that no longer serves a useful purpose. The category is broad. It includes large derelict objects, defunct satellites that have exhausted their propellant and ceased communicating, spent rocket upper stages left in orbit after delivering their payloads, and adapter rings or other large hardware released during deployment. It also includes the products of fragmentation, the cataloged fragments from past explosions and collisions, ranging from millimeter-scale flakes to half-meter pieces of structure. And it includes mission-related debris, items deliberately released during normal operations, such as lens covers, separation bolts, and pieces of multi-layer insulation that have torn loose from spacecraft over time.

The term excludes naturally occurring objects. Meteoroids, micrometeoroids, and the dust and gas of the interplanetary medium are not space debris in the technical sense, although they pose related hazards to spacecraft and are often grouped with debris under the combined term micrometeoroid and orbital debris, or MMOD. The technical distinction matters because different dynamics govern the populations and require different protective and operational responses. Natural micrometeoroids arrive predominantly from outside Earth's gravitational system, with characteristic velocities in the tens of kilometers per second; orbital debris is bound to Earth, with characteristic velocities of seven to fifteen kilometers per second relative to other Earth-orbiting objects. The flux from each population varies with altitude, and at most operational altitudes, the orbital debris flux now exceeds the natural meteoroid flux for objects larger than about 1 millimeter.

Technical literature divides the debris population into size categories that reflect both available tracking capabilities and the engineering responses available to spacecraft designers. The U.S. Space Surveillance Network reliably tracks objects larger than approximately 10 centimeters (3.94 inches) in low Earth orbit and approximately 1 meter (3 feet 3 inches) in geostationary orbit, with somewhat better resolution for objects with high radar cross-sections. ESA's space surveillance and tracking system, together with commercial providers such as LeoLabs, supplements U.S. tracking with additional sensors and somewhat different cataloging methodologies. The combined operational catalog for objects above the tracking threshold contains approximately 40,000 objects.

Categorization by Size

Large debris is conventionally defined as objects larger than ten centimeters. These objects are the primary concern for catastrophic collisions. A ten-centimeter (3.94-inch) object striking a satellite carries roughly the kinetic energy of a half-ton vehicle traveling at highway speed, concentrated into a small impact area at several miles per second relative to the target. No practical shielding can absorb such an impact. The only operational defense is conjunction screening, which predicts close approaches in advance and enables avoidance maneuvers. Each large object is also a potential source of thousands of smaller fragments if it is struck. Therefore, removing large derelict objects from the most densely populated orbital regions is one of the principal long-term debris-reduction strategies.

Medium debris, objects between 1 and 10 centimeters (0.39 to 3.94 inches) in size, constitute the most problematic population from an engineering standpoint. Objects in this range are too small for routine cataloging and tracking with current sensors, yet large enough to defeat any practical shielding. For example, a fragment less than half an inch (about one centimeter) across, made of aluminum and striking a satellite at orbital velocity, carries roughly the energy of a hand grenade. The estimated population of medium-sized debris in low Earth orbit, derived from radar measurements and statistical modeling, is approximately 1.2 million objects. Spacecraft designers cannot avoid these fragments operationally and cannot shield against them; the only mitigation strategy that addresses this population is to reduce the rate at which they are generated. The medium-debris population is, in this sense, the technical center of the debris problem: large enough to cause catastrophic damage, too small to track, and growing.

Small debris, defined as objects smaller than one centimeter (half an inch), is by far the most numerous category, with population estimates exceeding 140 million. Individual small fragments will not destroy a spacecraft outright, but they cause significant cumulative damage. Solar arrays, optical surfaces, thermal control coatings, and external sensors all degrade under continuous bombardment by small particles. The Long Duration Exposure Facility, a passive satellite retrieved by the Space Shuttle in 1990 after nearly six years in low Earth orbit, returned with an enormous number of small impacts that revealed both the density and the chemical composition of the small-debris population. The facility returned with more than 30,000 documented impacts across its 57 exposed experiments and trays. Analysis confirmed that at 476 kilometers altitude, the orbital debris flux for objects above 0.2 millimeters already exceeded the natural meteoroid flux by a factor of two to four, depending on the size range. This finding was the first large-scale empirical proof that debris, not nature, had become the dominant hazard for operational satellites in low Earth orbit. Subsequent shuttle missions, the windows of the International Space Station, and the surfaces of multiple research satellites have provided continuing data. The damage from small debris is rarely catastrophic on a per-impact basis. Still, it is the principal mechanism by which surfaces in orbit are progressively roughened, pitted, and eventually compromised.

The size categories are not perfectly distinct, and individual fragments span a continuous distribution. The conventional cutoffs, one millimeter, one centimeter, and ten centimeters, reflect breakpoints in observation and engineering response, not natural breaks in the debris population itself. A complete debris environment model must integrate over all sizes, with each size's contribution determined by the population density, the relative-velocity distribution, and the engineering significance of impacts at that size. Modern environment models, including NASA's ORDEM and ESA's MASTER, perform this integration explicitly, producing flux predictions tailored to specific spacecraft designs and operational orbits.

Categorization by Origin

An alternative classification organizes the debris population by how the objects were generated. Mission-related debris consists of objects deliberately released during normal spacecraft operations. This category includes spent rocket stages, payload adapter rings, deployment springs, lens covers ejected from optical sensors, fasteners released during separation events, and similar items. Mission-related debris is, in principle, the easiest category to address. Design changes can eliminate releases, and operational practices can ensure that necessary releases occur in orbits where decay is rapid. The debris-mitigation guidelines published by the IADC and adopted by most space agencies emphasize eliminating mission-related debris as a primary objective. Compliance has improved dramatically over the past two decades, and the contribution of new mission-related debris to the population has fallen sharply relative to fragmentation events. However, the historical population from earlier decades remains.

Collision-generated debris consists of fragments produced when two objects in orbit collide. The 2009 Iridium-Cosmos collision is a canonical example: two intact satellites, approximately 700 kilograms (1,543 pounds) and 950 kilograms (2,094 pounds), collided at approximately 11.7

kilometers per second (7.3 miles per second), producing approximately 1,800 cataloged fragments and an estimated population of much smaller debris numbering in the tens of thousands. Collision-generated debris is the most concerning category from a long-term sustainability standpoint because each fragment is itself a collision risk that can produce further fragments, a dynamic described by the Kessler syndrome and examined in detail in Chapter 4.

Explosion-generated debris consists of fragments produced when objects in orbit experience internal failures that cause disintegration. The most common cause is residual propellant in spent rocket upper stages. Pressurized propellant tanks subjected to thermal cycling and material degradation have, on numerous occasions, exploded years or decades after their original missions. The Delta upper-stage explosion problem of the 1970s and 1980s was the classic case, and modern passivation procedures, including venting residual propellant after spacecraft separation, were developed specifically to prevent such events. Battery explosions are a related cause, though less common. The 220 in-orbit fragmentation events tracked by ESA over the past two decades include a considerable proportion of explosion events, and the rate of new explosions has fallen significantly as passivation has become standard practice.

Anti-satellite test debris is a distinct subcategory of fragmentation debris that warrants separate consideration because of its strategic context and outsized contribution to the historical population. The Chinese anti-satellite test on January 11, 2007, against the Fengyun-1C weather satellite at an altitude of approximately 537 miles (865 kilometers) produced more than 3,500 cataloged fragments. The Russian anti-satellite test on November 15, 2021, against the defunct Cosmos 1408 satellite at an altitude of approximately 480 kilometers (298 miles) produced more than 1,500 cataloged fragments. These events are not accidents. They are the product of deliberate strategic decisions by national governments, made with full knowledge of the debris consequences, and they account for a large share of the total cataloged debris in low Earth orbit.

Degraded materials constitute a less dramatic but persistent category. Spacecraft surfaces, multi-layer insulation, paint, adhesives, and composite structures degrade in the orbital environment due to atomic oxygen, ultraviolet radiation, and thermal cycling. Particles are released continuously over a satellite's operational lifetime, contributing to the small-debris background. Aluminum oxide particles released during solid rocket motor firings also contribute. None of these sources alone dominates the debris environment, but their cumulative effect over thousands of missions is substantial.

The Physics of Hypervelocity Impact

Space debris poses severe risks to spacecraft because of the high relative velocities at which orbital objects move. An object in low Earth orbit travels at approximately 7.7 kilometers per second (4.8 miles per second), or about 28,000 kilometers per hour (17,398 miles per hour). Two objects in different orbital planes may approach each other at relative velocities exceeding 14 kilometers per second (8.7 miles per second). At those speeds, the kinetic energy of even small fragments is extraordinary. The standard expression for kinetic energy, one-half the mass times the velocity squared, makes this clear: doubling the velocity quadruples the energy, and orbital velocities are far higher than any encountered in terrestrial impacts.

A simple calculation illustrates the magnitudes involved. A one-gram fragment moving at 10 kilometers per second (6.2 miles per second) relative to a target carries 50 kilojoules of kinetic energy, equivalent to the energy of a 10-kilogram (22-pound) bowling ball thrown at 100 meters per second (224 miles per hour). A 100-gram fragment at the same velocity carries 5 megajoules, equivalent to the kinetic energy of a small car at highway speed. A 1-kilogram (2.2-pound) fragment carries 50 megajoules. These energies are concentrated, at the moment of impact, into volumes on the order of the projectile itself, with consequences that no conventional structural analogy adequately captures.

Hypervelocity impact, the term used in the engineering literature for impacts at velocities above approximately 3 kilometers per second (1.9 miles per second), behaves differently from conventional impact. Below the hypervelocity threshold, the projectile and target deform in ways consistent with conventional ballistics, and shielding can be designed to absorb kinetic energy through material strength and yielding. Above the threshold, the impact energy is high enough that both the projectile and the affected portion of the target temporarily behave as fluids. The projectile fragments completely; the target material in the impact region vaporizes, melts, or shatters. The damage volume in the target greatly exceeds the projectile's volume, and the damage extends well beyond the line of impact through the cloud of debris generated by the impact itself.

The Whipple shield, invented by astronomer Fred Whipple in 1947 and extensively refined for spacecraft applications, exploits this physics. A thin outer layer is positioned at a standoff distance from the spacecraft pressure hull. An incoming projectile strikes the outer layer at hypervelocity, and the resulting energy transfer disintegrates and disperses the projectile into a cloud of much smaller particles. The cloud expands across the standoff gap and strikes the inner hull over a much larger area, distributing enough energy that the inner hull can absorb the impact through conventional structural response. Modified Whipple shields with multiple layers, or with intermediate fabric layers such as Nextel or Kevlar, provide adequate protection against debris up to approximately one centimeter (half an inch) in diameter at characteristic orbital velocities. The walls of the International Space Station incorporate Whipple shielding throughout the crew-occupied modules and across most external surfaces.

Above the protective threshold, no practical shielding works. A two-centimeter (three-quarters of an inch) aluminum fragment cannot be stopped by any shield design that fits within the mass budget of an operational spacecraft. Therefore, defense against larger fragments must be operational rather than structural: the fragment must be tracked, the conjunction must be predicted, and the spacecraft must maneuver out of the threat's path. The full chain of capabilities that support this defense, including tracking, cataloging, prediction, decision-making, and maneuver execution, is examined in detail in Chapter 9.

Why the Numbers Matter

The lower bound for routine tracking, approximately ten centimeters (four inches) in low Earth orbit, has direct operational consequences. Objects above this threshold can be cataloged, their future trajectories predicted, and screened against active satellite trajectories to identify close approaches. Operators can then perform avoidance maneuvers, propellant permitting, when conjunction probabilities exceed acceptable thresholds. Objects below the threshold cannot be tracked individually, and their statistical distributions must be inferred from radar measurements of the population as a whole and from impact damage on the surfaces of returned objects.

The medium debris population, the 1.2 million objects between approximately 1 and 10 centimeters (.5 inches to 4 inches), is the critical gap. These objects can defeat shielding but cannot be tracked. Spacecraft designers cannot avoid them or armor against them. The only mitigation is to reduce the population itself, which means preventing the events that generate it. This is the operational reason the debris policy emphasizes eliminating fragmentation events as aggressively as it does. Each major fragmentation event adds thousands of trackable fragments and tens of thousands of medium fragments to the population, and the medium fragments persist as a permanent operational hazard for the lifetime of the affected altitude band.

The composition of the tracked population is instructive. According to ESA's 2025 figures, approximately 220 in-orbit fragmentation events have been recorded over the past two decades, and more than 3,000 trackable fragments were added to the catalog in 2024 alone. Roughly two-thirds of cataloged objects originate from a relatively small number of identifiable events: deliberate anti-satellite weapon tests, unintended collisions, and propellant-tank explosions. The implication is important. The space debris problem is not a diffuse byproduct of routine spaceflight. It is dominated by a manageable number of discrete events that are, in principle, preventable. A properly designed mitigation policy addresses the highest-leverage causes first.

Radar measurements of the small-debris population, conducted by ground-based phased-array radars and supplemented by impact-damage measurements on returned spacecraft surfaces, indicate that the small-debris flux varies by factors of three to ten across the principal altitude bands of low Earth orbit. The flux is highest near 800 kilometers (497 miles), where long-lived fragments from the Fengyun-1C and Iridium-Cosmos events are concentrated, and lower at higher altitudes, where atmospheric drag rapidly clears fragments. The flux distribution shapes the design of any spacecraft operating in low Earth orbit: shielding must be sized to the flux at the operational altitude, and the orientation of shielded surfaces must reflect the expected impact direction, which in turn depends on the orbital geometry of the dominant debris populations.

A Three-Part Problem

The space debris challenge can be understood as three interlocking problems. The first concerns the existing population: the objects already in orbit, accumulated over sixty-eight years of launches. This population includes large, defunct satellites and rocket bodies that pose an outsized risk because each, if struck, would generate thousands of new fragments. The second problem is the rate of new debris generation: the small but persistent flow of fragments from explosions, collisions, and operational releases that adds to the population each year, even when no serious events occur. The third problem is the trajectory of future activity: the rapid expansion of large satellite constellations, the increasing diversity of spacefaring nations and commercial operators, and the consequent rise in launch traffic and orbital density.

The three problems require distinct responses. The large-object population can be reduced through active debris removal, guided by international organizations such as the United Nations Office for Outer Space Affairs (UNOOSA) and by treaties such as the Outer Space Treaty. These missions are technically demanding and have only recently begun to move beyond the laboratory; Chapter 11 examines the current state of the field and the prospects for scaling. The rate of new debris generation is addressed primarily through mitigation: design standards that prevent on-orbit explosions, operational practices that minimize debris release, and disposal requirements that move spacecraft out of crowded regions at the end of life. Chapters 10, 7, and 8 examine the regulatory frameworks, including international guidelines and national policies, that codify these responses. The trajectory of future activity is shaped by policy: domestic licensing rules, international guidelines, traffic coordination, and the slow construction of governance frameworks adequate to a multipolar, increasingly commercial spacefaring world.

No single response solves the problem. Mitigation alone cannot remove debris already present. Removal alone cannot keep pace with the new generation if mitigation fails. Policy alone moves no mass. The integrated response the situation demands draws on engineering, operations, law, economics, and international diplomacy in roughly equal measure. This book treats each of those domains in turn, aiming to give readers a working understanding of how the elements fit together and where the principal leverage points lie.

The Stakes for Modern Civilization

Space debris can seem like a niche concern, a technical problem of interest only to satellite operators and aerospace engineers. The reality is that the satellite infrastructure now in orbit underwrites much of contemporary life. Global Positioning System satellites and their European, Russian, and Chinese counterparts provide the precise timing on which financial markets, electrical grids, and telecommunications networks depend, as well as the navigation services that have reshaped logistics, agriculture, and personal mobility. Weather satellites provide the observational data that feed every modern forecast and have measurably reduced the death toll from severe storms over the past forty years. Earth observation satellites monitor crops, deforestation, sea ice, and emissions. Communications satellites carry voice, data, and broadcast traffic to regions that terrestrial infrastructure cannot reach. Scientific satellites have transformed astronomy, planetary science, and the study of Earth's climate.

The dependence runs deeper than most users recognize. The timing signal broadcast by GPS satellites synchronizes cellular networks that route mobile calls, trading systems that execute orders on financial exchanges, and synchrophasor measurements (time-synchronized measurements of electrical voltage and current waveforms used to monitor and analyze power grid conditions in real time), enabling electrical grid operators to monitor power flow. The Federal Reserve's wire transfer system, the New York Stock Exchange's matching engine, and the synchronized timestamping of every credit card transaction in the world all depend, directly or indirectly, on the same satellite network that provides driving directions on a smartphone. The European Galileo, Russian GLONASS, and Chinese BeiDou constellations provide system-level redundancy, but their dependence on satellite-based timing is structural and difficult to replace.

Earth observation satellites have transformed industries in ways the public rarely understands. Modern agriculture uses satellite imagery for variable-rate fertilizer application, irrigation planning, and harvest scheduling, yielding measurable productivity gains that have helped feed a growing global population. Insurance companies use satellite data to verify damage claims after natural disasters and to underwrite agricultural risk. Logistics companies use satellite imagery to monitor port congestion, infrastructure conditions, and supply chain disruptions. Environmental agencies use satellite data to monitor air quality, deforestation rates, and the retreat of glaciers and ice sheets. Each of these applications depends on the continuous availability of satellite data; an orbital environment that disrupts that availability would have economic consequences far beyond the satellite industry itself.

A degraded orbital environment threatens all of this. The threat is not a single catastrophic event suddenly turning off the world's satellites; it is more insidious. As debris density rises, satellites must perform more avoidance maneuvers, consuming propellant that would otherwise extend their operational lives. Insurance premiums rise, increasing the cost of every commercial mission. New satellite designs must include heavier shielding or more redundant systems, increasing mass and reducing capability per dollar. The economics of doing business in orbit gradually deteriorate. At some threshold of density, the risk-adjusted cost of operating in a particular orbit exceeds the value of the services provided from that orbit, and the orbit is effectively abandoned. The question facing the spacefaring world is not whether such thresholds exist, but where they are and how soon they will be reached.

The Concept of Sustainability in Orbit

The phrase "space sustainability" entered serious policy discussion only in the past two decades. Its central proposition is that the orbital environment, like the atmosphere and the oceans, is a finite resource vulnerable to depletion from cumulative human activity. The United Nations Committee on the Peaceful Uses of Outer Space adopted the Guidelines for the Long-term Sustainability of Outer Space Activities in 2019, formalizing a framework that nations, agencies, and operators are expected to apply voluntarily. The guidelines address the regulatory environment, the safety of space operations, international cooperation, and scientific and technical cooperation. They do not have the force of a treaty, and compliance is uneven, but they represent the most coherent international statement of what responsible behavior in orbit entails.

Sustainability in this context is not aspirational language. It is a measurable property of the orbital environment that can, with appropriate engineering and policy, be maintained and, without those measures, will not be. The 2009 collision between the operational Iridium 33 communications satellite and the defunct Russian Cosmos 2251 produced approximately 1,800 cataloged fragments, many of which remain in orbit at altitudes around 789 kilometers (490 miles), well within the heart of the most heavily used low Earth orbit shells. The 2007 Chinese anti-satellite test against the Fengyun-1C weather satellite produced more than 3,500 cataloged fragments at 865 kilometers (537 miles), of which approximately 2,800 are still tracked. These two events alone account for a substantial fraction of all current debris in low Earth orbit. They are practical illustrations of what happens when sustainability is treated as somebody else's problem.

The sustainability framework has been operationalized through several complementary mechanisms. The Inter-Agency Space Debris Coordination Committee's Mitigation Guidelines, originally published in 2002 and revised in 2021, provide the technical baseline for spacecraft design and operations. National regulators, particularly in the United States and Europe, have begun to translate these guidelines into binding requirements through licensing rules. The Space Sustainability Rating program, originally launched by the World Economic Forum and now run by the École Polytechnique Fédérale de Lausanne in Switzerland, provides operators with a structured way to evaluate and publicly disclose the sustainability characteristics of their missions. None of these mechanisms is sufficient on its own; together, they constitute a developing infrastructure for the long-term governance of the orbital environment.

Mega-Constellations and the Acceleration of Change

The 2010s and 2020s have brought a structural change to the orbital environment more consequential than any single fragmentation event. The deployment of large satellite constellations, first by SpaceX with

Starlink beginning in 2019, and subsequently by OneWeb, Amazon's Project Kuiper (renamed Amazon Leo in November 2025), and a series of Chinese constellations, including Guowang and Qianfan, has roughly doubled the population of active satellites in orbit over about five years. As of mid-2026, Starlink operates approximately 10,400 to 10,500 active satellites; OneWeb operates approximately 648 satellites under the Eutelsat Group following a 2023 merger; and Amazon Leo is in active deployment, with approximately 302 production satellites in orbit by April 2026. Several thousand additional satellites are projected for the late 2020s.

The constellation era brings both new risks and new opportunities. The risks are obvious: more satellites mean more potential collisions, more avoidance maneuvers, and greater demand on the conjunction-screening infrastructure. The opportunities are less immediately visible but real. Modern constellation satellites are designed with disposal in mind, typically at altitudes where atmospheric decay will clear them within five to seven years after mission completion. They are equipped with propulsion adequate for collision avoidance and end-of-life disposal. They are operated by entities with strong commercial incentives to maintain a usable orbital environment. And the sheer scale of these operations has driven investment in autonomous collision avoidance, precise tracking, and traffic coordination, benefiting the entire community.

Whether the constellation era is, on balance, a positive or negative development for the debris environment will depend on how reliably operators meet their disposal commitments and how effectively the regulatory framework adapts to the scale of operations. The experience of the past five years has been mixed. The total number of in-orbit fragmentation events has risen in proportion to the number of satellites, suggesting that mitigation discipline has not kept pace. At the same time, the failure of even a small fraction of constellation satellites to perform end-of-life disposal could, over a decade or two, significantly increase the long-term debris population. The community is, in this respect, in the middle of a large and consequential experiment whose outcome will be visible only over the coming decades.

International Efforts to Address Space Debris

The international response to the debris problem has progressed slowly but with measurable results. The Inter-Agency Space Debris Coordination Committee, founded in 1993, brings together the major national space agencies to coordinate on technical matters and develop common mitigation standards. The committee's Space Debris Mitigation Guidelines, published in 2002 and revised in 2021, provide the technical framework adopted by most national agencies and many commercial operators. The United Nations Committee on the Peaceful Uses of Outer Space adopted aligned guidelines in 2007 and the Long-term Sustainability Guidelines in 2019.

Treaty-level instruments addressing space activities, including the 1967 Outer Space Treaty, the 1972 Liability Convention, and the 1976 Registration Convention, predate the explicit recognition of debris as a sustainability problem and do not directly regulate debris generation. They, however, establish the legal framework within which subsequent debris-related instruments operate. The Outer Space Treaty's prohibition on national appropriation of celestial bodies, its requirement of state responsibility for the activities of national operators, and its preservation of space for peaceful purposes provide the foundation for contemporary debris policy. The 1972 Liability Convention establishes that launching states are absolutely liable for damage caused by their space objects on the surface of the Earth and fault-liable for damage caused in space. The Registration Convention requires states to maintain registries of objects launched into space and to provide that information to the United Nations.

Domestic regulatory frameworks have evolved more rapidly than international instruments. In September 2022, the U.S. Federal Communications Commission adopted a five-year disposal rule for low Earth orbit satellites licensed to provide service in the United States, replacing the previous twenty-five-year guideline with a more aggressive standard. The European Space Agency has progressively tightened its internal mission requirements through successive revisions of its Space Debris Mitigation Policy. The United Kingdom, France, Japan, and several other spacefaring states have adopted similar standards through their national licensing authorities. The pattern is one of progressive tightening at the national level, with international consensus following in a delayed manner. Chapter 7 examines the international instruments in detail; Chapter 8 examines the national frameworks.

The Artemis Accords, opened for signature in October 2020 and signed by more than fifty nations as of mid-2026, extend the international framework beyond Earth orbit to govern lunar and cislunar activity. The accords establish principles for transparency, interoperability, the registration of space objects, the release of scientific data, the preservation of historic sites, and the responsible disposal of orbital debris. Although the accords are not a binding treaty, they represent the most significant addition to the international space governance framework since the 2019 LTS Guidelines.

Emerging Solutions for Debris Mitigation

The technical responses to the debris problem fall into three broad categories: prevention, removal, and protection. Prevention encompasses the design and operational practices that prevent the generation of new debris. Modern satellites are equipped with propulsion systems capable of performing end-of-life disposal maneuvers, are designed to minimize the release of mission-related debris, and include passivation procedures that vent residual propellant and discharge batteries before the spacecraft becomes inactive. The trend in spacecraft design has steadily shifted toward more demanding mitigation standards, and the cost of compliance has declined as the industry has gained experience with relevant technologies.

Active debris removal, the direct capture and de-orbit of existing derelict objects, has progressed from laboratory studies to operational demonstration over the past five years. Astroscale's ELSA-d mission demonstrated rendezvous and capture of a cooperative client object in 2021. ClearSpace-1, originally scheduled to remove a VESPA payload adapter, was retargeted in April 2024 to the smaller PROBA-1 satellite after the original target was struck by debris and deemed unusable; the launch is now planned for 2028. The PRELUDE mission, announced in January 2026 and targeting a 2027 launch, will further demonstrate active removal capabilities. None of these missions, individually, materially changes the debris population. Their importance lies in establishing technical capabilities and regulatory precedents on which a future scaled removal capability could be built.

Protection technologies, Whipple shields, redundant systems, and operational procedures provide the last line of defense against debris that prevention and removal cannot address. The walls of the International Space Station incorporate sophisticated shielding designs, validated by decades of impact data. Modern commercial satellites apply Whipple shielding selectively to critical components and in directions of expected impact. Onboard collision avoidance, increasingly automated, allows satellites to maneuver out of the path of identified threats with minimal operator intervention. The combination of shielding and avoidance reduces, but does not eliminate, the operational risk posed by debris.

A fourth, emerging category is environmental management, recognizing that the debris problem extends beyond orbital congestion to include the atmospheric and astronomical effects of space activity. Reentries deposit metals into the upper atmosphere; recent measurements have detected aluminum, lithium, and copper from reentered satellites in stratospheric aerosols at altitudes around 18 kilometers (11 miles). Large constellations contribute to light pollution that affects ground-based astronomy, with implications for both research and the cultural value of the night sky. Chapter 6 examines the reentry questions; Chapter 15 examines the astronomical and environmental effects in detail.

A Map of the Book

The chapters that follow develop these themes in order. Chapter 2 traces the historical evolution of the orbital debris problem, from the first Cold War launches through the era of mega-constellations. It identifies when the problem was first recognized, when warnings were ignored, and when events ultimately forced the issue onto the international agenda. Chapter 3 examines the orbital regimes themselves, including low Earth orbit, medium Earth orbit, geostationary orbit, and highly elliptical orbits used for specialized missions, as well as the orbital mechanics that govern them and the distinctive vulnerabilities of each. Chapter 4 introduces the Kessler syndrome, the cascade scenario first described by Donald Kessler and Burton Cour-Palais in 1978, and assesses how close the present environment is to the conditions Kessler described.

Chapter 5 examines the principal fragmentation events of the satellite era as case studies, drawing out the technical and political lessons from each. Chapter 6 addresses reentry hazards and the atmospheric effects of reentering spacecraft, a topic that has become increasingly urgent in recent years. Chapters 7 and 8 examine the international and national legal frameworks that govern space activity and limit debris generation. Chapter 9 examines the technical infrastructure for tracking, space situational awareness, and space traffic management. Chapter 10 examines mitigation as a discipline of design and operations. Chapter 11 examines active debris removal and in-orbit servicing. Chapter 12 examines artificial intelligence, autonomy, and emerging technologies. Chapter 13 examines the economics, insurance, and sustainability ratings that translate environmental conditions into commercial decisions. Chapter 14 examines human spaceflight and the specific concerns of crewed missions. Chapter 15 examines astronomy, atmospheric pollution, and the broader environmental commons. Chapter 16 examines the cislunar environment and the path forward.

What This Book Is Not

A few clarifications are helpful. This is not a technical manual. The treatment of orbital mechanics, hypervelocity impact physics, and spacecraft engineering is detailed enough to support the argument but does not replace the textbooks and standards practitioners consult. This is not a polemic. The data support the argument that the orbital environment is at risk. Still, they permit a range of policy responses, and the book seeks to present the principal options fairly rather than advocate for one. This is not a forecast. Predictions about the trajectory of the debris environment depend on assumptions about future launch traffic, mitigation compliance, and removal capacity that no one can know with certainty. Where the book offers projections, it identifies the assumptions and sources.

The book claims that the orbital debris problem is well-defined, that responses to it are technically tractable, that the principal obstacles are economic and political rather than scientific, and that the choices made in the next decade by governments, agencies, and operators will shape the orbital environment that the rest of this century inherits. The book's point is to make those choices visible and to give the general reader, the student, the policymaker, and the practitioner a common reference for the discussion that follows.

Chapter Summary

The space debris environment, the byproduct of nearly seven decades of orbital activity, now contains roughly 40,000 tracked objects and at least a hundred million smaller fragments, almost all traveling at velocities sufficient to destroy any satellite they strike. The environment poses real and growing risks to satellite infrastructure that supports navigation, communications, weather forecasting, scientific observation, and national security. The problem has three components: the existing population, the rate of new debris generation, and the trajectory of future activity. Addressing it requires an integrated response across engineering, operations, policy, and international cooperation. The chapters that follow develop each of these elements in turn, providing readers with a complete and current understanding of where the problem stands, how it arose, and what is being done about it.

Chapter 2. The Historical Evolution of Orbital Debris

Introduction

The history of space debris is inseparable from the history of space exploration. Every gram of orbital debris was, at some point, deliberately placed in orbit by a human launch operation. The accumulation of debris over six and a half decades reflects the cumulative decisions of national space programs, commercial operators, military planners, and individual mission designers about what to launch, where to operate, and what to leave behind. Some of those decisions were made with full awareness of the long-term consequences; many more were made without that awareness. The result is the orbital environment we live in today.

This chapter traces that history in roughly chronological order, with attention to the moments when the problem was first recognized, the events that drove it forward, and the responses that gradually emerged. The treatment is necessarily selective: the full history of space activity is too vast to summarize in a single chapter, and the focus here is specifically on the events and decisions that shaped the debris environment. Readers interested in the broader history of space exploration will find many excellent treatments in the literature; the references at the end of this book identify several.

The Origins: Sputnik and the Space Race

The space age began as a competition. The Soviet Union's launch of Sputnik 1 on October 4, 1957, was understood by the United States and its allies less as a scientific milestone than as a strategic shock. The same R-7 rocket that placed an 83.6-kilogram (184-pound) aluminum sphere in orbit could, in principle, deliver a thermonuclear warhead anywhere on Earth. The American response, beginning with the formation of the Advanced Research Projects Agency in February 1958 and the establishment of the National Aeronautics and Space Administration in October of the same year, was framed in the same competitive terms. The first successful American satellite, Explorer 1, reached orbit on January 31, 1958, and discovered the Van Allen radiation belts.

By the end of 1960, the United States and the Soviet Union had together launched dozens of satellites, including the first navigation, weather, and reconnaissance satellites. The pace of activity through the 1960s was extraordinary. The Apollo, Soyuz, Mariner, Venera, Surveyor, Lunar Orbiter, Mercury, Gemini, and Voskhod programs, along with a steadily expanding constellation of communications, weather, and reconnaissance satellites, all flew during this decade. Each launch placed at least one upper stage in orbit alongside the spacecraft, and each spacecraft eventually became debris when its mission ended. Mission planning of the era universally treated end-of-life disposal as an afterthought. Spacecraft were designed to function for as long as possible. They were left in their operational orbits when they ceased to function, on the assumption that orbital decay would eventually clear them or that the volume of space was too large for accumulation to matter.

This assumption was reasonable in the early years but increasingly indefensible thereafter. Decay indeed clears objects from low altitudes, but with a strong altitude dependence: an object at 400 kilometers (249 miles) may decay within months to years. By comparison, an object at 800 kilometers (497 miles) may persist for centuries, and one at 1,000 kilometers (621 miles) or more may persist for thousands of years. The orbital regions that proved most useful for operational satellites, such as the bands between 600 and 1,200 kilometers (373 and 746 miles) used for sun-synchronous Earth observation, or the geostationary belt at 35,786 kilometers (22,236 miles) used for telecommunications and weather imaging, were precisely the regions where decay was slowest and accumulation most pronounced. The early space community did not generally think in these terms; the concept of orbital crowding belonged to a future they did not anticipate.

The Soviet Pioneering Role

The Soviet space program of the 1950s and 1960s, under the leadership of Sergei Korolev and his successors, achieved a remarkable series of firsts. Sputnik 2, launched on November 3, 1957, carried the dog Laika and became the first spacecraft to orbit a living animal. The Luna program returned the first images of the lunar far side in 1959 with Luna 3 and achieved the first soft lunar landing in 1966 with Luna 9. The Venera program made the first successful landings on Venus, beginning with Venera 7 in 1970, and operated under conditions, including surface temperatures above 460 degrees Celsius (860 Fahrenheit) and atmospheric pressures ninety times that of Earth, that no American spacecraft has yet attempted. The Vostok program, beginning with Yuri Gagarin's flight on April 12, 1961, established Soviet primacy in human spaceflight for several years. Each of these programs left hardware in orbit, on the lunar surface, on Venus, and increasingly throughout the solar system.

The Soviet program was marked by a pace of activity that often outpaced the United States. Through the 1960s, Soviet launches outnumbered American launches by a substantial margin, although the average sophistication of American payloads, particularly in communications and reconnaissance, was higher. The cumulative effect on the orbital environment was a Soviet-dominant population of upper stages and defunct satellites in low Earth orbit, with a particular concentration at the inclinations and altitudes preferred by the Soviet program. Many of these objects are still in orbit today, six decades later. Soviet rocket bodies from the 1960s and 1970s are among the largest single objects in the current debris catalog, and the Inter-Agency Space Debris Coordination Committee's working lists of priority targets for active debris removal have consistently identified Soviet-era stages as among the highest-leverage candidates.

One specific legacy of the Soviet program deserves separate mention. The Radar Ocean Reconnaissance Satellite program, known in the West as RORSAT, deployed approximately 33 nuclear-powered radar satellites between 1967 and 1988, each carrying a nuclear reactor fueled by highly enriched uranium. When these satellites reached the end of their missions, standard practice was to fire a small rocket that transferred the reactor core to a higher storage orbit, typically around 800 to 950 kilometers, intended to keep radioactive material out of the atmosphere for several hundred years until the reactor products decayed to safe levels. Approximately 16 of these reactor cores remain in those storage orbits today, slowly degrading as the sodium-potassium coolant loops have been leaking since the end of their missions. The leaking coolant has produced a cloud of small, frozen droplets in the same altitude bands, creating a category of small but highly reflective debris that is difficult to track with conventional radar. The RORSAT reactors represent a form of long-lived debris hazard with no parallel in the Western space program, and their eventual management is an unresolved legacy of the Cold War space competition.

The American Response and the Apollo Era

The American response to early Soviet successes evolved through several phases. The Mercury program, running from 1961 to 1963, proved that humans could survive launch, orbit, and reentry. The Gemini program, running from 1965 to 1966, refined the techniques of rendezvous, docking, and extravehicular activity that would be required for lunar missions. The Apollo program, beginning with the Apollo 1 fire in January 1967 and culminating in the Apollo 11 lunar landing on July 20, 1969, established American primacy in deep-space exploration. Six successful crewed lunar landings between 1969 and 1972 returned 382 kilograms (842 pounds) of lunar samples to Earth and laid the scientific foundation for the modern understanding of the Moon's origin and history.

From a debris standpoint, the Apollo era was distinctive because much of its hardware did not remain in Earth orbit. The Saturn V upper stages used for translunar injection either escaped Earth orbit on lunar trajectories or deliberately impacted the lunar surface to simulate seismic events for instruments left by earlier missions. The lunar modules, command modules, and service modules met similar fates. The Skylab space station, launched in 1973 and reentered in 1979, was the largest Apollo-era contribution to the Earth-orbital debris environment. Still, its low inclination and modest altitude meant it decayed within six years of its abandonment. The July 1979 reentry of Skylab, with debris falling across Western Australia, was the first widely publicized example of an uncontrolled reentry of a large spacecraft; the Western Australian town of Esperance famously fined NASA for littering, an event that has remained part of space-history folklore.

The Apollo era contributed to the orbital debris environment indirectly through the dramatic increase in launch traffic it occasioned, the deployment of the first communications and reconnaissance satellite networks, and the establishment of operational practices that left spent stages in orbit and released mission-related debris during deployment, practices that became routine. These practices were not specific to American programs; the Soviet program operated similarly, and the smaller programs that emerged later, French, British, Japanese, and Chinese, generally adopted the practices of the established programs. The cumulative effect was that the operational baseline of the 1960s and 1970s built debris-generating practices into the foundations of every subsequent space program.

The Cold War Era and Military Uses of Space

The Cold War shaped the orbital environment more decisively than any single technology. The first reconnaissance satellites, Corona for the United States, Zenit for the Soviet Union, established the use of orbit for strategic intelligence collection. The first navigation satellites, Transit for the United States, Tsiklon for the Soviet Union, proved that submarine-launched ballistic missiles could be guided to global accuracy. The first early-warning satellites established the use of orbit to detect ballistic missile launches. By the end of the 1960s, each of these systems was an integral part of the strategic nuclear architecture of one or both superpowers.

The military significance of orbit changed both the volume of activity and the character of debris generation. The first deliberate anti-satellite weapons emerged in this period. The Soviet Union conducted a long series of co-orbital anti-satellite intercepts between 1968 and 1982 under the Istrebitel Sputnikov program, each of which produced fragments at altitudes between 500 and 1,000 kilometers (311 and 621 miles). The American direct-ascent ASM-135 test on September 13, 1985, in which a missile launched from an F-15 destroyed the Solwind P78-1 satellite at

555 kilometers (345 miles), produced approximately 285 cataloged fragments. The strategic justification for these tests was deterrence, demonstrating that one superpower could threaten the other's satellite assets. Still, the consequences of the debris were deemed acceptable because the strategic benefit was deemed paramount.

The Strategic Defense Initiative of the 1980s, popularly known as Star Wars, accelerated debris-related concerns in a different direction. The proposed deployment of large numbers of intercept platforms in low Earth orbit, combined with the contemplated use of kinetic-kill vehicles and directed-energy weapons, raised the prospect of conflict scenarios in which orbital debris generation would be massive. Although the most ambitious elements of the program were never deployed, the technical analyses that supported it pushed the debris community to consider scenarios well beyond the routine accumulation of past decades. The end of the Cold War in 1989-1991 reduced the immediate prospect of large-scale orbital conflict. Still, it did not eliminate the underlying capability, and the strategic competition that returned in the 2010s and 2020s would revive many of the same concerns.

Anti-Satellite Weapons and the Debris Legacy

The pattern of anti-satellite testing changed in the post-Cold War era but did not end. The Chinese anti-satellite test of January 11, 2007, in which a direct-ascent kinetic kill vehicle destroyed the Fengyun-1C weather satellite at approximately 865 kilometers (537 miles) altitude, produced more than 3,500 cataloged fragments and an estimated population of much smaller fragments in the tens of thousands. Because of the high altitude, the fragments are expected to persist for decades to centuries. The Fengyun-1C event remains the single largest contributor to the current low Earth orbit debris population. Roughly 2,800 fragments from the test remain cataloged more than nineteen years after the event.

The American response to the Chinese test came in February 2008, when the United States destroyed its own malfunctioning USA-193 reconnaissance satellite as part of Operation Burnt Frost. The intercept was conducted at an altitude of approximately 247 kilometers (153 miles), deliberately low, to ensure that fragments would decay rapidly. Approximately 175 fragments were cataloged, and most reentered within months. The American test was justified on the basis that residual hydrazine fuel aboard the satellite would survive uncontrolled reentry and pose a hazard on the ground; whatever the merits of that justification, the choice of low altitude reflected an awareness of the debris consequences that the Chinese test had not displayed. The contrast became a standard example in the debris policy literature of how strategic objectives can be achieved with widely varying environmental footprints.

Subsequent anti-satellite tests have continued the pattern. India's Mission Shakti test on March 27, 2019, was conducted at an altitude of approximately 282 kilometers (175 miles), deliberately low, and produced approximately 130 cataloged fragments, most of which decayed within a year. Russia's anti-satellite test on November 15, 2021, against the defunct Cosmos 1408 satellite at approximately 480 kilometers (298 miles), produced more than 1,500 cataloged fragments and forced the crew of the International Space Station to take shelter in their return spacecraft as the fragment cloud passed nearby. The events bracket the range of contemporary state behavior: deliberate fragmentation at high altitude remains within the toolkit of certain national programs. In contrast, others have begun to test more responsibly or to refrain entirely.

Following the Russian test, the United States announced in April 2022 a unilateral commitment to refrain from destructive direct-ascent anti-satellite testing, and several other states have made similar commitments through United Nations resolutions. By 2024, more than thirty countries have endorsed the commitment, including all of the principal European spacefaring states, Japan, South Korea, Canada, and Australia. China and Russia have not endorsed it, although both have refrained from destructive testing since their last tests. The norm, while not universal and not legally binding, represents the most significant development in anti-satellite arms control in decades and a measurable improvement in protecting the orbital environment from deliberate degradation.

The Commercial and Collaborative Era

The 1990s brought structural changes to the orbital environment that the preceding decades had not. The end of the Cold War reduced the strategic competition that had driven much of the launch traffic. Still, it simultaneously opened the orbital environment to a wider range of national programs and to commercial operators in numbers the previous era had not seen. The deregulation of telecommunications in many countries and the maturation of geostationary communications technology produced a steadily increasing population of commercial communications satellites. The deployment of the first global navigation satellites for civilian use, GPS, achieved Initial Operational Capability in December 1993 and Full Operational Capability in April 1995, laying the foundation for a vast range of commercial applications to develop later.

The first major satellite constellations appeared in this period. The Iridium constellation, deployed beginning in 1997, placed 66 operational satellites in low Earth orbit at an altitude of approximately 780 kilometers (485 miles) to provide global mobile telephony. The Globalstar constellation, deployed beginning in 1998, placed 48 satellites at an altitude of approximately 1,400 kilometers (870 miles) for similar purposes. Both constellations encountered severe commercial difficulties; the underlying business case proved fragile in the face of the rapid expansion of cellular networks, but the technical infrastructure they established laid the groundwork for the much larger constellations of the 2010s and 2020s. The Iridium constellation, in particular, provided the satellite that would later collide with Cosmos 2251 in the most consequential accidental collision of the satellite era, an event examined in detail in Chapter 5.

The commercial era also brought new participants into the spacefaring community. The European Space Agency, founded in 1975, came of age operationally in the 1990s with the development of the Ariane launch vehicle family, the construction of a robust European satellite manufacturing industry, and the establishment of independent European capabilities in Earth observation, navigation, and scientific exploration. The Japanese space program, beginning with the founding of the National Space Development Agency in 1969 and merging into JAXA in 2003, demonstrated Japanese launch capability with the H-II family of rockets and produced notable scientific missions. The Indian space program, beginning under Vikram Sarabhai in the 1960s and later consolidated under ISRO, became operationally prominent in the 1990s with the development of the Polar Satellite Launch Vehicle and the deployment of an extensive Earth observation constellation. The Chinese space program, often less visible internationally, expanded steadily through the same period and emerged in the 2000s as a fully capable spacefaring power.

The International Space Station

The International Space Station, conceived in 1984 as Space Station Freedom and later expanded to include Russian, European, Japanese, and Canadian participation in the early 1990s, has been the largest single human presence in orbit since assembly began in 1998. The station has been continuously crewed since November 2, 2000, longer than any other crewed spacecraft in history. Its orbital altitude has ranged from about 370 to 460 kilometers (230 to 286 miles) over its operational life, with an inclination of 51.6 degrees, enabling access from Russian launch sites at Baikonur and American launch sites at Cape Canaveral. The station orbits at about 28,000 kilometers per hour (17,398 miles per hour) and completes one orbit every 90 minutes.

The station is the most extensively studied example of long-term human presence in a debris-rich orbital environment. The station has performed dozens of debris-avoidance maneuvers over its operational life, including several prompted by fragments from anti-satellite tests. The station's external surfaces have been continuously monitored for impact damage. Multiple windows have been replaced after small debris impacts caused chipping and cracking. The station's emergency procedures include shelter protocols for the crew when fragment clouds pass nearby, most prominently highlighted during the November 2021 fragmentation of Cosmos 1408. The station is currently planned to remain operational until 2030, after which it will be deorbited via a controlled reentry using a NASA-procured deorbit vehicle developed by SpaceX.

Beyond its specific contributions to debris science, the ISS is a model of international collaboration in space that has proven durable amid periods of substantial geopolitical strain. The station has continued to operate amid deteriorating US-Russian relations since 2014 and again after 2022, with day-to-day technical cooperation persisting even when broader political relations have been hostile. Whether the model is reproducible in the post-ISS era, when the principal spacefaring partners may operate separate stations rather than a single shared facility, remains an open question for the coming decade.

The Modern Era of Mega-Constellations

The 2010s brought a structural change to the orbital environment that was more transformative than any single fragmentation event. The deployment of large satellite constellations, first by SpaceX with Starlink beginning in 2019, and subsequently by OneWeb, Amazon's Project Kuiper (renamed Amazon Leo in November 2025), and a series of Chinese constellations including Guowang and Qianfan, has roughly doubled the population of active satellites in orbit over about five years. As of mid-2026, Starlink operates approximately 10,400 to 10,500 active satellites; OneWeb operates approximately 630 to 650 satellites under the Eutelsat Group following a 2023 merger; and Amazon Leo is in active deployment, with approximately 302 production satellites in orbit by April 2026. Several thousand additional satellites are projected for the late 2020s.

The technical characteristics of these constellations differ in ways that matter for the debris environment. Starlink satellites operate predominantly at altitudes between 540 and 570 kilometers (336 and 354 miles), where atmospheric drag clears nonfunctional satellites within five to seven years. OneWeb satellites operate at approximately 1,200 kilometers (746 miles), where natural decay timescales are much longer, so end-of-life disposal must be performed with propulsion. Amazon Leo satellites are being deployed across multiple shells between approximately 590 and 630 kilometers (367 and 391 miles). The Chinese constellations are reportedly being deployed at varying altitudes, including some at higher altitudes that would imply more challenging disposal requirements.

The constellation business model is more sensitive to the debris environment than earlier satellite businesses. A single geostationary communications satellite, operated for 15 years and serving customers across an entire continent, can absorb higher avoidance-maneuver costs without seriously affecting unit economics. A constellation satellite, operated for five to seven years and serving customers within a small ground footprint at any given moment, has much tighter margins. Each avoidance maneuver consumes propellant that would otherwise extend operational life. Each fragmentation event in a nearby altitude band increases the rate of avoidance maneuvers across the constellation. The cumulative effect of even modest debris growth, multiplied across thousands of satellites and decades of operation, becomes a material business risk.

The constellation era has also begun to shape regulatory expectations in ways the previous generation of operations did not. The U.S. Federal Communications Commission's adoption of a five-year disposal rule in September 2022 was made possible, in part, by demonstrating that constellation satellites operating at approximately 550 kilometers (342 miles) could comply with the new rule without extensive additional engineering. The European licensing framework has moved in similar directions. Whether the pattern of regulatory tightening continues and whether the operational practices of constellation operators remain acceptable to regulators in other jurisdictions will be among the central questions of the coming decade.

The First Recognition of the Problem

The technical recognition that orbital debris would become a self-sustaining problem dates to a 1978 paper in the Journal of Geophysical Research. Donald Kessler, then a researcher at NASA's Johnson Space Center, and Burton Cour-Palais, a colleague specializing in hypervelocity impact, published "Collision Frequency of Artificial Satellites: The Creation of a Debris Belt." Drawing on techniques originally developed to study asteroid evolution, Kessler and Cour-Palais showed that as the number of artificial objects in low Earth orbit increased, the probability of collisions would rise dramatically. Each collision would produce fragments, each of which would become a collision risk. At sufficient density, the process would become self-sustaining: collisions would generate fragments faster than the atmosphere removed them, and the orbital region would, over decades, fill with a belt of debris analogous to the asteroid belt.

The paper's projections were striking. Kessler and Cour-Palais estimated that, at late-1970s launch rates and assuming no mitigation, the cascade could become destabilizing within a century. They noted, however, that fragments from upper-stage explosions and earlier anti-satellite tests might already be present in sufficient numbers to shorten this interval. The paper was published without internal NASA approval and circulated rapidly within the international debris community. Within a few years, NASA established the Orbital Debris Program Office at the Johnson Space Center, with Kessler as its first head, and laid the analytical foundation for the modern debris problem.

The Delta upper-stage explosions of the 1970s and early 1980s gave Kessler's argument its first operational weight. Between 1973 and 1981, at least nine Delta Thorad Agena and Delta upper stages exploded in orbit, most at altitudes between 800 and 1,500 kilometers. Each had been left in orbit with residual UDMH and nitrogen tetroxide propellant. Thermal cycling over months to years allowed pressure to build until the tanks failed. NASA's investigation found that a single passivation burn, venting residual propellant through the engines at the end of the mission, would eliminate the hazard. The first standardized passivation procedures were written in the early 1980s. That a procedural change costing almost nothing in mass or cost could prevent the single largest category of fragmentation events was, in retrospect, one of the clearest lessons the debris community has ever been handed. Chapter 4 examines the Kessler argument and its modern descendants in detail.

Recognition, however, did not translate quickly into action. The 1980s saw both the continuation of debris-generating activity and the slow accumulation of measurements confirming the analytical predictions. The first Long Duration Exposure Facility, a passive satellite designed to study the effects of long-term exposure to the space environment, was recovered by the Space Shuttle in 1990 after nearly six years in orbit. It was found to have extensive impact damage from both micrometeoroids and orbital debris. The data from LDEF, combined with similar exposure data from returning spacecraft surfaces, established conclusively that the debris flux at low altitudes already exceeded the natural meteoroid flux. The transition from theoretical possibility to measured reality occurred over approximately 15 years, from Kessler's 1978 paper to the consolidated empirical picture of the early 1990s.

The Cold War Legacy in Detail

Several aspects of the Cold War approach to space contributed disproportionately to the debris environment that exists today. The first was the routine practice of leaving rocket upper stages in orbit. Many of these stages contained residual propellant in pressurized tanks. Over time, thermal cycling, propellant decomposition, and tank corrosion led to a series of explosions, each generating hundreds or thousands of fragments. The Delta upper-stage explosion problem, in particular, was identified in the 1970s and largely solved through passivation procedures that vented residual propellant after spacecraft separation. Still, the population of unpassivated stages launched in the preceding decades continued to explode for years afterward. Upper-stage explosions of this kind dominated the catalog of fragmentation events from the 1970s to the 1980s.

The second contribution was deliberate anti-satellite testing, discussed earlier. The third contribution was a category of operational releases that would today be considered unacceptable but were standard in the early decades. Lens caps were ejected from optical sensors. Bolt fragments were released during stage separation. Cooling fluid was vented from the radiators. Each release was small, but the cumulative effect over thousands of missions was a population of small debris that no tracking system could catalog and no shielding could fully address. When mitigation discipline eventually appeared, it would target this category of release as aggressively as the larger fragmentation events.

A fourth, less-discussed contribution was the practice of detonating spent satellites and rocket bodies as a form of housekeeping. The Soviet Union conducted multiple deliberate breakups of older satellites in the 1970s and 1980s, in some cases as tests of anti-satellite capabilities and in others as straightforward disposal operations. The American program engaged in similar practices on a smaller scale. The cumulative debris from these deliberate breakups was considerable, and many of the resulting fragments remain in orbit today. The practice ended in both programs by the late 1980s as the consequences of debris became better understood, but its legacy persists. The catalog of historical fragmentation events compiled by NASA and ESA includes dozens of events that were neither accidents nor military operations, but rather deliberate disposals carried out under the operational assumptions of an earlier era.

The First Mitigation Frameworks

The transition from observation to action accelerated in the 1990s. The Inter-Agency Space Debris Coordination Committee (IADC) was established in 1993 as a forum for national space agencies, initially NASA, ESA, the Russian Space Agency, and the Japanese space agency, with subsequent expansion, to coordinate research and develop common technical standards. NASA published its first internal Safety Standard for Limiting Orbital Debris in 1995, establishing requirements for passivation, mission-related debris, and end-of-life disposal that would influence future international standards. The IADC Space Debris Mitigation Guidelines were published in 2002 and, with subsequent revisions, remain the most widely cited technical framework for mitigation. The United Nations Committee on the Peaceful Uses of Outer Space endorsed an adapted version of the IADC guidelines in 2007.

The IADC guidelines codified what is sometimes called the twenty-five-year rule: the recommendation that spacecraft operating in low Earth orbit be removed from the protected region within twenty-five years of mission completion. The rule was a compromise. It was lenient enough to be technically and economically achievable for the satellite designs of the early 2000s, yet strict enough to limit the long-term accumulation of debris in the most heavily used orbital regions. Compliance with the twenty-five-year rule has been notoriously uneven. ESA's analyses of compliance have repeatedly shown that fewer than half of all satellites reaching the end of life in low Earth orbit are disposed of in accordance with the guidelines. However, compliance has improved measurably over the past decade as enforcement mechanisms have tightened.

The most significant policy development in recent years has been the reduction of the disposal timeline. In September 2022, the United States Federal Communications Commission adopted a five-year rule for low Earth orbit satellites licensed to provide service in the United States, replacing the previous twenty-five-year guideline with a more aggressive standard. The rule applies to satellites disposed of after September 2024 and represents one of the first instances of a national regulator imposing requirements stricter than the international consensus. Other regulators have begun to adopt similar approaches, and these policies have significantly influenced the design of new satellite constellations, particularly those operating at altitudes where natural orbital decay alone would not satisfy the updated disposal requirements. Chapters 7 and 8 examine the international and national regulatory frameworks in detail.

The Era of the Large Events

Three events between 2007 and 2009 transformed the debris environment more than any policy change during that period. The first was the Chinese anti-satellite test on January 11, 2007, against the Fengyun-1C weather satellite at an altitude of approximately 865 kilometers (537 miles), producing more than 3,500 cataloged fragments. The second was the American anti-satellite test on February 20, 2008, against the malfunctioning USA-193 satellite at an altitude of approximately 247 kilometers (153 miles), producing approximately 175 cataloged fragments, most of which decayed within months. The third was the accidental collision on February 10, 2009, between the operational Iridium 33 communications satellite and the defunct Russian Cosmos 2251 at an altitude of approximately 789 kilometers (490 miles) over northern Siberia, producing approximately 1,800 cataloged fragments.

The Iridium-Cosmos collision warrants particular attention because it was the first known accidental hypervelocity collision between two intact satellites and because its causes were operational rather than strategic. The two satellites had been tracked individually for years. The U.S. Space Surveillance Network had identified the close approach in advance. Still, the predicted miss distance was reported as several hundred meters, well above the threshold that would have prompted Iridium to perform an avoidance maneuver. The collision occurred because the conjunction analysis lacked sufficient precision in the cross-track direction; the predicted miss distance was off by an amount that, in retrospect, fell within the prediction's uncertainty. The lesson was that the operational and regulatory infrastructure to prevent such collisions systematically had not yet matured. Conjunction screening services, which compare the predicted positions of all tracked objects and warn operators of close approaches, expanded in the years that followed; the modern infrastructure is examined in Chapter 9.

Recent Developments and the Trajectory Ahead

The most recent additions to the orbital environment have come from a mix of sources. The November 2021 Cosmos 1408 anti-satellite test, discussed above, was the most notable deliberate fragmentation event of the past decade. Several critical commercial fragmentation events have occurred since 2018, mostly involving the explosions of older Chinese and Russian rocket upper stages. As a result, the rate of fragmentation events has been higher over the past five years than in any comparable period since the 2007-2009 era of large events.

On the positive side, the past five years have also seen the first operational missions explicitly devoted to debris removal, the deployment of autonomous collision avoidance across commercial constellations, tighter regulatory disposal requirements in multiple key jurisdictions, and the development of a mature commercial space situational awareness sector. The Astroscale ELSA-d mission in 2021, the planned ClearSpace-1 mission scheduled for 2028, and the PRELUDE mission targeting 2027 mark the first wave of operational debris-removal capability. Whether these capabilities will be developed at the scale and pace required by the debris environment is the central technical question of the coming decade.

The historical record suggests that debris-generating activity has occurred in episodes, periods of relative restraint punctuated by sudden bursts of fragmentation. These episodes are not random. Each reflects decisions by states or operators to prioritize other considerations over the consequences of debris. The capacity of the international system to discourage such decisions, while accommodating the legitimate strategic and commercial interests of spacefaring actors, is the central political question of the coming decade. The chapters that follow examine the technical, legal, and institutional infrastructure on which that capacity depends.

Chapter Summary

The orbital debris environment today is the product of nearly seven decades of cumulative decisions by national space programs, military planners, and commercial operators. The early decades treated the orbital environment as effectively infinite and regarded end-of-life disposal as an afterthought. Recognition that this approach was unsustainable emerged gradually, first through the analytical work of Kessler and Cour-Palais in 1978, then through empirical confirmation in the 1980s and early 1990s, and finally through the development of mitigation frameworks by the

IADC, NASA, ESA, and other agencies in the 1990s and 2000s. The major fragmentation events of 2007-2009 and 2021 proved that the historical pattern of debris generation had not been broken. They also catalyzed substantial improvements in mitigation discipline, conjunction screening, and regulatory tightening. The continued deployment of large constellations will mark the coming decade, the development of operational debris-removal capability, and the testing of the international system's ability to constrain deliberate debris generation. The chapters that follow develop each of these elements in detail.

Chapter 3. Orbital Regimes and Mechanics

Why Orbits Matter

An orbit is not a place. It is a relationship between an object's altitude, its velocity, and the gravitational field of the body it circles. Two satellites at very different points in space may share an orbit if they have the same orbital parameters. Two satellites that are physically close to each other may belong to entirely different orbital regimes if their inclinations or eccentricities differ. Understanding the debris environment requires recognizing that the term "low Earth orbit" denotes a band of altitudes, not a single shell, and that the densities, lifetimes, and risks within that band vary with altitude, inclination, and operational practices.

The principal regimes used by operational spacecraft are low Earth orbit, medium Earth orbit, geostationary orbit, the highly elliptical orbits used for specialized communications and scientific missions, and the graveyard orbits used for the disposal of geostationary satellites. Each regime is associated with a characteristic set of missions, a characteristic debris environment, and a characteristic vulnerability to the cascade processes described in the next chapter. This chapter treats each regime in turn, following a review of the orbital mechanics governing them.

The Mechanics of Orbit

The behavior of objects in orbit is governed by Kepler's three laws of planetary motion, originally formulated for the planets of the solar system but applicable to any body moving under the gravitational influence of a much more massive central body. Johannes Kepler derived these laws between 1609 and 1619 from Tycho Brahe's observational data, two generations before Newton would provide their dynamical foundation. The laws establish the geometric and temporal properties of any orbit and remain the starting point for every introductory treatment of orbital mechanics.

The first law, the law of ellipses, states that an orbit is an ellipse with the central body at one focus. A circular orbit is a special case of an elliptical orbit with zero eccentricity. The orbit of a typical low Earth orbit satellite, such as those in the Starlink constellation, is nearly circular, deliberately so, because circular orbits maintain a constant altitude and therefore consistent operational performance. Highly elliptical orbits, with eccentricities of 0.5 or greater, are used for specialized purposes where the geometric properties of the ellipse provide operational advantages. The Russian Molniya orbit, with a perigee of approximately 600 kilometers (373 miles) and an apogee of approximately 39,000 kilometers (24,233 miles), is the canonical example.

The second law, the law of equal areas, states that a line connecting the orbiting body to the central body sweeps out equal areas in equal times. This means an orbiting object moves faster when it is closer to the central body and slower when it is farther away. In practical terms, a satellite in an elliptical orbit spends more time at apogee, the highest point of its orbit, than at perigee. Communications satellites in Molniya orbits exploit this property: a satellite in a Molniya orbit spends approximately eight hours of each twelve-hour orbit at high altitude over the northern hemisphere, providing extended communications coverage to high-latitude regions where geostationary satellites are difficult to use.

The third law, the harmonic law, states that the square of an orbit's period is proportional to the cube of its semi-major axis. In practical terms, larger orbits take longer to complete, and the relationship is geometric rather than linear. The constant of proportionality depends on the mass of the central body. For Earth, the relationship implies that an orbit at 400 kilometers (249 miles) altitude has a period of approximately 90 minutes; an orbit at 800 kilometers (497 miles) has a period of approximately 100 minutes; an orbit at 20,000 kilometers (12,427 miles) has a period of approximately 12 hours; and an orbit at 35,786 kilometers (22,236 miles), the geostationary altitude, has a period of exactly one sidereal day (the time the Earth takes to complete one full rotation relative to the distant stars, rather than relative to the Sun). The harmonic law is the basis for calculating the geostationary altitude and for predicting the orbital periods of all spacecraft.

Newton's Law of Gravitation

Newton's law of universal gravitation, published in 1687 in the Principia Mathematica, provided the dynamical foundation for Kepler's laws. The gravitational force between two bodies is proportional to the product of their masses and inversely proportional to the square of the distance between them. The constant of proportionality, the gravitational constant G, is approximately 6.674×10^{-11} $N \cdot m^2/kg^2$. For an object in orbit around a vastly more massive central body, the force simplifies to a simple expression: the gravitational acceleration at any point depends only on the distance from the central body's center.

The orbital velocity required to maintain a circular orbit at a given altitude can be derived from Newton's law and equals the square root of the gravitational parameter divided by the orbital radius. The gravitational parameter, the product of G and Earth's mass, is approximately 3.986×10^{14} m^3/s^2. For low Earth orbit at an altitude of 400 kilometers (249 miles), the orbital radius is approximately 6,778 kilometers (4,212 miles) (Earth's radius plus the altitude), and the orbital velocity is approximately 7.67 kilometers per second (4.8 miles per second). For a geostationary orbit at an altitude of 35,786 kilometers (22,236 miles), the orbital radius is approximately 42,164 kilometers (26,199 miles), and the orbital velocity is approximately 3.07 kilometers per second (1.9 miles per second).

Escape velocity, the speed required for a projectile to escape Earth's gravitational influence entirely, is approximately 11.2 kilometers per second (7 miles per second) at Earth's surface and decreases with altitude. At a given altitude, escape velocity is about 1.414 times the circular orbital velocity at that altitude. Spacecraft destined for the Moon, the planets, or interplanetary space must reach escape velocity, typically through staged rocket maneuvers or transfer orbits. The mechanics of these maneuvers, including Hohmann transfers, bi-elliptic transfers, and gravity assists, form the foundation of mission design, and the cumulative debris consequences of mission design choices over six decades have shaped the orbital environment that exists today.

The Six Orbital Elements

A complete description of an orbit requires six parameters, traditionally called the orbital elements. The semi-major axis defines the size of the orbit, equal to half the longest diameter of the ellipse. The eccentricity defines the shape, with zero corresponding to a circle and values approaching one to highly elongated ellipses. The inclination is the angle between the orbital plane and Earth's equatorial plane, ranging from zero degrees for an equatorial orbit to 180 degrees for a retrograde equatorial orbit, with 90 degrees corresponding to a polar orbit. The right ascension of the ascending node describes the orientation of the orbital plane around Earth's axis, measured from the vernal equinox direction. The argument of perigee describes the orientation of the elliptical orbit within its plane, measured from the ascending node to the point of closest approach. The true anomaly describes the spacecraft's position along its orbit at a specified epoch.

These six parameters fully specify a Keplerian orbit, that is, an orbit influenced only by the gravitational pull of a central point mass. Additional effects perturb real orbits, including Earth's oblateness, atmospheric drag, lunisolar gravitational forces, and solar radiation pressure. The orbital elements drift over time under the influence of these perturbations. The drift dynamics are central to several aspects of debris management, as discussed below. Operational tracking systems maintain time-tagged orbital elements for every cataloged object, updating them as new measurements become available and propagating them forward in time to predict positions for conjunction screening.

Inclination and Orbital Plane

An orbit's inclination has consequences that extend well beyond the geometric orientation of the orbital plane. Launch site latitude determines the minimum inclination achievable from a given site without a costly plane-change maneuver. The Russian launch site at Baikonur, located at 46 degrees north latitude, can launch directly into orbits with an inclination of 46 degrees or higher. The American launch site at Cape Canaveral, at 28 degrees north, can reach lower-inclination orbits more efficiently, and most American geostationary satellites take advantage of this geometry. The European launch site at Kourou, in French Guiana, at 5 degrees north, is closer to the equator than any other major launch site, providing the most efficient access to geostationary orbit and explaining why ESA selected the location.

The inclination of an orbit also determines which regions of Earth's surface a satellite can observe or communicate with. A satellite in a low-inclination orbit can communicate efficiently with ground stations near the equator but never passes over high latitudes. A satellite in a polar orbit, with an inclination near 90 degrees, eventually passes over every point on Earth's surface as the planet rotates beneath the orbital plane. Earth observation missions that require global coverage typically use polar or near-polar orbits for this reason. The historical concentration of Earth observation, weather, and reconnaissance missions in polar and near-polar orbits has produced one of the densest debris populations in low Earth orbit, with implications discussed below.

The relative velocity at which two objects approach each other depends critically on the angle between their orbital planes. Two objects in identical orbital planes but at slightly different altitudes may approach each other at relative velocities of only tens of meters per second; an impact at such a velocity is unlikely to cause significant fragmentation. Two objects in retrograde orbits, one moving eastward and the other westward, approach each other at twice their orbital velocity, more than 15 kilometers per second (9.3 miles per second); an impact at such a velocity is catastrophic for both spacecraft. Therefore, the inclination distribution of the satellite and debris populations matters as much as the altitude distribution. The operational consequences of this geometry are examined throughout the rest of this book.

Perturbations and Long-Term Stability

A real orbit is not a perfect Keplerian ellipse. It is perturbed by several effects that cause its parameters to drift over time. The largest perturbation in low Earth orbit, the forces in space that slowly change a spacecraft's orbit, is Earth's oblateness, conventionally described by the J2 coefficient. The J2 coefficient is a mathematical value that represents the effect of Earth's equatorial bulge on its gravitational field, causing satellite orbits to change slowly over time. Earth is not a perfect sphere;

it is slightly flattened at the poles and bulges at the equator, with a difference of approximately 21 kilometers (13 miles) between equatorial and polar radii. The resulting gravitational field is not perfectly spherical, and this deviation produces a torque on inclined orbits, causing the orbital plane to precess around Earth's axis.

The rate of J2 precession depends on altitude and inclination. In typical low Earth orbits, the precession rate is on the order of several degrees per day. Sun-synchronous orbits deliberately exploit J2 precession, choosing inclinations near 98 degrees so that the orbital plane precesses at the same rate as Earth's motion around the Sun. As a result, the satellite crosses the equator at the same local solar time on every pass. This property is invaluable for Earth observation: imagery of the same scene at the same local time enables comparisons across days, seasons, and years that would otherwise be confounded by varying illumination. The historical concentration of Earth observation missions at sun-synchronous altitudes, particularly between 600 and 800 kilometers (373 and 497 miles), is a direct consequence of this exploitation.

Atmospheric drag is the dominant perturbation at low altitudes. The residual atmosphere at orbital altitudes is dominated by atomic oxygen at 400 kilometers (249 miles) and shifts to lighter species at higher altitudes. Density varies dramatically with solar activity: during solar maximum, when the Sun's ultraviolet flux heats and expands the upper atmosphere, the density at 400 kilometers (249 miles) can be two to three times higher than during solar minimum. Satellites at this altitude lose altitude at rates ranging from about 50 meters (164 feet) per day during solar minimum to several hundred meters per day during solar maximum. Reboost maneuvers, which use propulsion to raise the orbit, are required at intervals ranging from weeks to months for satellites operating at these altitudes. The International Space Station performs reboosts approximately monthly when fully active, with timing depending on solar conditions, station mass, and visiting vehicle traffic.

The solar activity cycle, with its approximately eleven-year periodicity, has direct implications for debris management. The current solar cycle, Cycle 25, peaked around 2024-2025, producing the most intense solar maximum since Cycle 23 in the early 2000s. The increased atmospheric drag during solar maximum has measurably accelerated the decay of debris in low Earth orbit, with several thousand objects deorbiting in 2024 alone. However, the same enhanced drag has increased the operational burden on low-altitude satellites, requiring more frequent reboosts and consuming additional propellant. The interaction between solar activity and the debris environment is one of the more complex aspects of long-term debris modeling.

Lunisolar perturbations, the gravitational influence of the Moon and Sun, affect higher orbits more than lower ones. In the geostationary orbit, the combined effect produces a long-term drift in inclination, with satellites that are not actively maintained accumulating inclination at approximately 0.85 degrees per year. Station-keeping maneuvers correct this drift; satellites that exhaust their station-keeping propellant before mission end fall outside the geostationary belt and become candidates for graveyard disposal. Solar radiation pressure, the small but persistent force exerted by sunlight on illuminated surfaces, affects highly elliptical orbits and objects with high area-to-mass ratios. Collectively, these perturbations mean that real orbits are stable on short timescales but evolve over years to decades, and that the long-term debris environment is shaped by perturbation dynamics as much as by the original placement of objects in orbit.

Low Earth Orbit

Low Earth orbit, conventionally defined as the region between approximately 200 and 2,000 kilometers (124 and 1,243 miles) altitude, is the most heavily used orbital regime and contains the largest populations of both active satellites and debris. The International Space Station orbits at approximately 408 kilometers (254 miles) altitude with an inclination of 51.6 degrees. Most Earth observation satellites operate between 500 and 800 kilometers (311 and 497 miles) above Earth. Most communications constellations operate between 500 and 1,200 kilometers (311 and 746 miles). The dense satellite shells of the Starlink, OneWeb, and Amazon Leo constellations are concentrated in this band.

The orbital velocity at these altitudes is approximately 7.7 kilometers per second (4.8 miles per second), decreasing slightly with altitude. An orbit at 400 kilometers (249 miles) takes about 90 minutes to complete; an orbit at 1,200 kilometers (746 miles) takes about 110 minutes. The relative velocity between two objects in different low Earth orbits depends on the geometry of the encounter. However, relative velocities of 10 to 14 kilometers per second (6.2 to 8.7 miles per second) are characteristic of the most concerning collision geometries, those between objects in different inclination planes.

The atmospheric environment varies dramatically across low Earth orbit. At 400 kilometers (249 miles), residual atmospheric drag is significant enough that satellites must perform regular reboosts to maintain altitude. The decay timescale for an inactive satellite at this altitude is on the order of months to a few years, depending on solar activity and the satellite's ballistic coefficient. At 800 kilometers (497 miles), the timescale stretches to decades. At 1,000 kilometers (621 miles), it stretches to a century or more. The Sun's eleven-year activity cycle modulates the upper atmosphere's density, with periods of high solar activity producing greater drag and shorter satellite lifetimes than periods of low activity. This dependence makes the precise prediction of long-term decay times difficult. It gives the upper-altitude bands of low Earth orbit an effective permanence that has shaped both historical debris accumulation and current mitigation policy.

The density of debris in low Earth orbit is concentrated in two altitude bands. The first is around 800 to 900 kilometers (497 to 559 miles), where long-lived fragments from the Fengyun-1C anti-satellite test and the Iridium-Cosmos collision are concentrated. The second is around 1,400 to 1,500 kilometers (870 to 932 miles), where historical sun-synchronous orbit constellations have left a long-lasting population. Below approximately 600 kilometers (373 miles), atmospheric drag clears debris on timescales short enough that the density remains moderate even with high launch traffic. The deployment of commercial constellations at altitudes between 540 and 570 kilometers (336 and 354 miles) reflects, in part, recognition that this band offers a useful combination of operational performance and natural debris cleanup.

Sun-Synchronous and Polar Orbits

A subset of low Earth orbit warrants particular attention because of its operational importance and outsized contribution to the debris environment. Sun-synchronous orbits are inclined at approximately 98 degrees. They are chosen so that the orbital plane slowly rotates at the same rate as Earth's motion around the Sun, resulting in a satellite in a sun-synchronous orbit crossing the equator at the same local solar time on every pass. This property is invaluable for Earth observation: imagery of the same scene at the same local time enables comparisons across days, seasons, and years that would otherwise be confounded by varying illumination.

The historical concentration of Earth observation missions, weather satellites, and intelligence-gathering platforms in sun-synchronous orbits, particularly at altitudes between 600 and 800 kilometers (373 and 497 miles), has produced an unusually dense and long-lived debris population in these orbits. The Fengyun-1C fragmentation occurred near a sun-synchronous altitude. Several major historical breakups have occurred in or near sun-synchronous orbits. The polar inclination of these orbits also means that conjunctions between objects in different sun-synchronous planes occur frequently, with relative velocities approaching the maximum possible in low Earth orbit. The sun-synchronous bands are, in effect, a worst-case region within low Earth orbit, combining high operational density, long debris lifetimes, and adversarial collision geometry. Modeling studies consistently identify these altitudes as the regions where the cascade processes described in the next chapter would first manifest.

The operational community has begun to respond to this concentration. Several recent Earth observation missions have selected altitudes below 600 kilometers (373 miles), where atmospheric drag clears debris more rapidly, and natural disposal is achievable without onboard propulsion. The trade-off is between coverage, with high-altitude orbits providing larger ground footprints, and sustainability, as low-altitude orbits clear themselves but require more satellites to provide equivalent coverage. Economics have shifted toward lower altitudes as launch costs have declined and the cost of regulatory non-compliance has risen. Still, the historical population at higher sun-synchronous altitudes remains a long-term hazard that no operational change can address directly.

Medium Earth Orbit

Medium Earth orbit spans the range from the top of low Earth orbit, approximately 2,000 kilometers (1,243 miles), to the bottom of geostationary orbit, approximately 35,786 kilometers (22,236 miles). The most important operational use of medium Earth orbit is for navigation satellite constellations: the American Global Positioning System at approximately 20,200 kilometers (12,552 miles), the Russian GLONASS at approximately 19,100 kilometers (11,868 miles), the European Galileo at approximately 23,200 kilometers (14,416 miles), and the Chinese BeiDou, which uses a mix of medium Earth orbit and geostationary orbit. A small number of communications satellites also operate at medium Earth orbit altitudes, including the SES O3b constellation at approximately 8,000 kilometers (4,971 miles).

Medium Earth orbit is relatively uncongested. The total population of active satellites is in the low hundreds. Atmospheric drag is negligible, and the band spans a vast volume, so spatial densities are far lower than in low Earth orbit. The debris environment is also relatively benign, in part because this orbital zone has been used almost exclusively by sophisticated state-sponsored programs that, overall, have observed mitigation guidelines. No severe fragmentation event has occurred there.

A complicating factor specific to medium Earth orbit is the Van Allen radiation belts, two concentric bands of trapped charged particles surrounding Earth at altitudes of roughly 1,000 to 6,000 kilometers and 13,000 to 60,000 kilometers. The inner belt is centered near the altitudes of the lower GPS and GLONASS constellations. The intense radiation environment in these belts accelerates the degradation of satellite electronics and solar cells, shortening effective mission lifetimes. It also means that any future debris-removal mission targeting objects at these altitudes must be designed to operate in an environment that rapidly degrades conventional spacecraft components. No removal vehicle has yet been designed or operated there, and the technical demands are far greater than those faced by low Earth orbit removal missions.

The principal vulnerabilities of medium Earth orbit differ from those of low Earth orbit. Because objects in medium Earth orbit do not decay naturally, any debris generated in this orbital region is essentially permanent on human timescales. A single critical fragmentation event could degrade this orbital region in ways that atmospheric cleanup could not reverse. The regime's strategic importance, as a source of global navigation services on which a wide range of civilian and military systems depend, also makes it a target of particular concern in any future anti-satellite scenario. The regime's relatively benign current state should not be mistaken for resilience; it primarily reflects the small number of objects placed there and the absence of the kinds of accidents and tests that have marked the lower altitudes.

Graveyard orbits used to dispose of medium Earth orbit satellites typically lie several hundred kilometers above operational orbits. GPS satellites, when retired, are disposed of at approximately 20,500 kilometers (12,738 miles) in a region intended to remain clear of the operational constellation. The disposal practices for GLONASS, Galileo, and BeiDou are similar, although the specific graveyard locations differ. Long-term management of these graveyard regions is an emerging concern. As the number of disposed navigation satellites grows, the prospect of collisions among them cannot be ruled out indefinitely.

Geostationary Orbit

Geostationary orbit is a circular, equatorial orbit at approximately 35,786 kilometers (22,236 miles) altitude, where the orbital period equals one sidereal day. A sidereal day is the time it takes Earth to complete one full rotation relative to the distant stars, about 4 minutes shorter than the normal 24-hour solar day used in everyday life. A satellite in this orbit appears to remain stationary above a fixed point on the equator, making the regime uniquely valuable for telecommunications and weather observation. A handful of widely viewed events from a geostationary perspective, such as imagery of full hemispheres delivered by GOES, Himawari, Meteosat, and similar weather satellites, and the broadcast of major sporting and political events to multiple continents, depend on the geometric properties of this orbit, and no other orbital regime offers a substitute.

The geostationary belt is a thin region. The orbital altitude tolerance for stationkeeping is about plus or minus 50 kilometers (31 miles), and the latitudinal tolerance is about plus or minus a few degrees. Approximately 600 active satellites occupy this volume, distributed around the equator at longitude slots assigned by the International Telecommunication Union. The slots themselves are an economically valuable resource, and the management of slot assignments under the ITU's regulations is one of the older and more institutionalized examples of international orbital governance. A geostationary slot at a useful longitude can be worth tens of millions of dollars in licensing value, and disputes over slot allocation have been a recurring theme in international space governance for decades.

Two characteristics of the geostationary orbit shape its debris environment. The first is that geostationary satellites do not decay. An object placed in geostationary orbit, or in an orbit close to it, remains there essentially forever on any practical timescale. The second is that the IADC and related guidelines specify a graveyard orbit, approximately 235 kilometers (146 miles) above the geostationary belt, into which satellites are expected to be moved at the end of their operational lives. The graveyard altitude is set to ensure that disposed satellites do not drift back into the operational region due to perturbations over the next century. Compliance with the graveyard requirement has historically been imperfect: many older satellites lacked the propellant to perform the disposal maneuver, and some operators failed to plan for it. However, compliance has improved markedly over the past two decades. Most modern geostationary satellites reach their graveyards successfully.

Geostationary orbit has not, to date, experienced a significant fragmentation event. Several breakups of geostationary satellites and apogee kick motors have occurred, but they have produced fragments at low rates and have not visibly degraded the regime. The principal long-term concern for geostationary orbit is the graveyard region itself, where the population of disposed satellites is now substantial and where future collisions among graveyard objects could, over decades, generate fragments that drift back into the operational region. Active management of the graveyard is an open problem in current debris policy, and the eventual development of in-orbit servicing and repositioning capabilities may prove essential to its long-term sustainability.

Highly Elliptical Orbits

Highly elliptical orbits are a specialized class used primarily for communications and scientific missions that require extended dwell time over high latitudes. The Russian Molniya orbit, with a perigee of approximately 600 kilometers (373 miles), an apogee of approximately 39,000 kilometers (24,233 miles), and an inclination of 63.4 degrees, was developed to provide communications coverage of the Soviet Union's high-latitude territory in an era before the advent of geostationary satellites with steerable beams. The Tundra orbit, with a similar geometry but a 24-hour period, has been used for related applications. Several scientific missions, including the Chandra X-ray Observatory and several heliophysics missions, also use highly elliptical orbits for their distinctive viewing and radiation properties.

The choice of inclination for the Molniya orbit reflects a specific orbital-mechanics consideration: at 63.4 degrees, the J2 perturbation does not cause the argument of perigee to drift. Hence, the apogee remains fixed over the high-latitude region the orbit is designed to serve. The mathematical condition that produces this stability, called the critical inclination, is a special case of the broader perturbation dynamics described above and illustrates how orbital design routinely exploits specific features of the perturbation environment.

The debris environment in highly elliptical orbits is comparatively understudied. These orbits cross multiple regimes, spending part of each period in low Earth orbit and part in or near geostationary altitudes. Cataloging and tracking objects in these orbits is more difficult than for circular orbits at standard altitudes. Fragmentation of objects in highly elliptical orbits can distribute debris across a wide range of altitudes, complicating the modeling of long-term consequences. As the use of these orbits is currently modest and likely to remain so, the debris consequences are limited; nonetheless, the regime is a reminder that the catalog of operational orbits extends well beyond the principal three regimes.

Cislunar and Beyond

The orbital regimes discussed above are all within Earth orbit. As the Artemis program and subsequent crewed lunar missions advance, the cislunar environment, the region of space between Earth and the Moon, including the Lagrange points where the gravitational forces of the two bodies balance, will become operationally important in ways that the previous decades of space activity have not anticipated. The Earth-Moon Lagrange points L1 and L2, located approximately 60,000 kilometers (37,282 miles) from the Moon, are particularly attractive for staging spacecraft, communications relays, and scientific instruments. Several missions have already flown through or near these locations; many more are planned for the coming decade.

The debris implications of cislunar activity are not yet well characterized. The dynamics of cislunar trajectories are dominated by the combined gravitational fields of Earth and the Moon, with substantial contributions from the Sun and solar radiation pressure. Objects in cislunar space do not experience atmospheric drag, though they may eventually escape the Earth-Moon system or impact either body, depending on their trajectories. The timescales of these natural processes typically range from years to centuries, longer than those in low Earth orbit but shorter than those in geostationary orbit.

The cislunar region poses regulatory and governance challenges that existing frameworks address only partially. The Outer Space Treaty applies in cislunar space as it does elsewhere. However, the technical guidelines developed by the IADC and other bodies were not designed with cislunar activity in mind. The Artemis Accords, opened for signature in October 2020, address some aspects of lunar surface activity but do not comprehensively address cislunar orbital debris. Developing appropriate regulatory frameworks for cislunar activity is one of the open challenges of the coming decade, and Chapter 16 returns to this topic in detail.

Implications for Mitigation and Removal

The regime-specific nature of the debris environment directly affects the design of mitigation and removal strategies. A disposal requirement appropriate at one altitude may be impractical at another. The FCC's five-year disposal rule is technically achievable for satellites operating below approximately 600 kilometers (373 miles), where atmospheric decay assists. At higher altitudes, however, it requires onboard propulsion and active deorbit maneuvers, with consequences for satellite mass, design complexity, and cost. The choice of priority targets for active debris removal depends on which orbital regions are most at risk of cascade and which individual objects, if removed, would produce the largest reduction in long-term collision probability. Modeling studies have repeatedly identified large derelict objects at altitudes between 700 and 1,000 kilometers (435 and 621 miles) as the highest-priority removal targets, and missions that have begun to demonstrate removal capability have selected their initial targets accordingly.

Orbital mechanics also constrain the operational envelope of debris removal missions. A removal vehicle launched into the target object's orbital plane can perform the rendezvous efficiently; one that must change planes must expend enormous amounts of propellant, raising mission cost. The economics of debris removal, therefore, favor mission architectures that target multiple objects in similar orbital planes rather than single-target missions. Several proposed removal mission concepts are organized around this principle, with single launches addressing several priority objects within a single sun-synchronous or near-sun-synchronous shell.

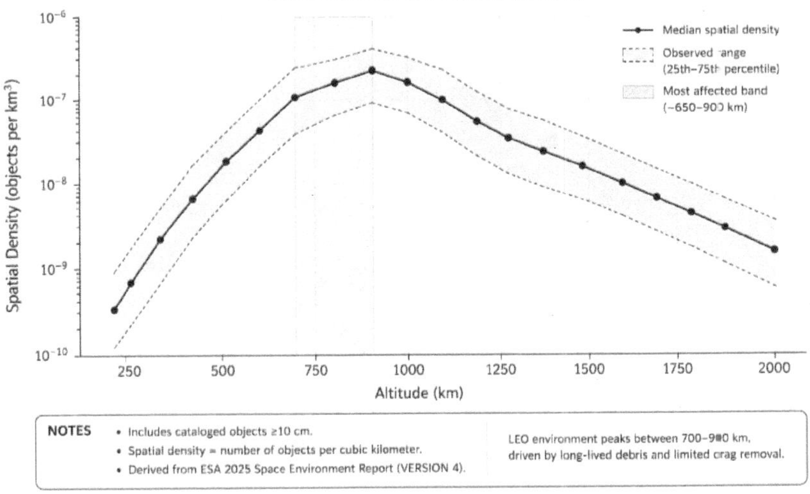

Figure 3.1. Spatial Density of Cataloged Objects in Low Earth Orbit by Altitude
(Based on ESA 2025 Space Environment Report)

Figure 3.2. Inclination Distribution of Cataloged Objects in Low Earth Orbit

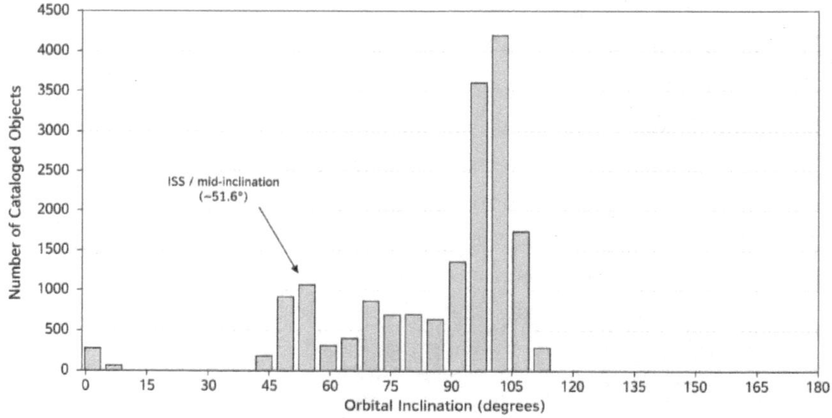

NOTES
- Includes cataloged objects ≥10 cm.
- Bins are 5-degree inclination intervals.
- Derived from ESA 2025 Space Environment Report (VERSION 4).

The distribution is bimodal, with a dominant population in sun-synchronous orbits (~90–100°) and a secondary population near the ISS inclination (~51.6°).

Figure 3.3. The Geostationary Belt: Operational Orbit, Graveyard Region, and Libration Nodes

(Earth and orbital distances to scale)

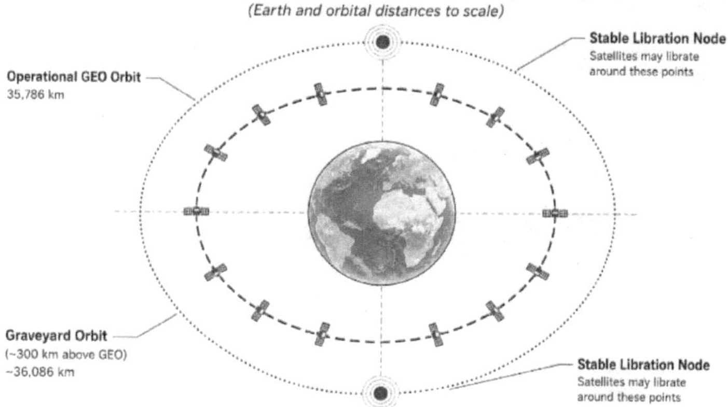

Operational GEO Orbit Altitude: 35,786 km Used by active satellites	**Graveyard Orbit** Altitude: ~36,086 km Used for retired satellites	**Stable Libration Nodes** Regions where gravitational forces allow satellites to "librate" or hover

Regime	Altitude Range Approximately	Orbital Period	Typical Use	Decay Time
Low Earth Orbit (LEO)	160–2,000 km 100-1,243 Miles	90–127 min	ISS, Earth observation, comms constellations	Years to centuries
Medium Earth Orbit (MEO)	2,000–35,786 km 1,243-22,236 Miles	2–24 hours	GPS, GLONASS, Galileo, BeiDou	Thousands of years
Geostationary Orbit (GEO)	35,786 km 22,236 Miles	23h 56m	Communications, weather, broadcast	Effectively permanent
Geostationary Transfer Orbit (GTO)	200-35,786 km 124-22,236 Miles	~10.5 hours	Transfer orbits to GEO	Decades to centuries
Highly Elliptical Orbit (HEO)	1,000- 40,000 km 620-24,850+Miles	Variable	Molniya, Tundra, science missions	Decades to centuries
Sun-Synchronous Orbit (SSO)	600–800 km 373-497 Miles	~98 min	Earth observation, reconnaissance	Decades to centuries

Table 3.1. Characteristics of Principal Orbital Regimes

Chapter Summary

Understanding the orbital regimes is the foundation for everything that follows. The mechanics of orbits, Kepler's laws, Newton's law of gravitation, the six orbital elements, and the perturbations that cause real orbits to drift over time help determine where spacecraft operate, how long debris persists, and which collisions are physically possible. The principal operational regimes, low Earth orbit, sun-synchronous orbit, medium Earth orbit, geostationary orbit, and highly elliptical orbit, each have distinct debris environments, distinct vulnerabilities to cascade, and distinct mitigation requirements. The cislunar region, which has become increasingly important in this decade, poses regulatory and technical challenges that existing frameworks address only partially. The remainder of the book treats the cascade dynamics that the regimes make possible, the events that have already produced large-scale debris, the policies and technologies that aim to limit further damage, and the question of how the orbital environment can be managed for the long term.

Chapter 4. The Kessler Syndrome and Cascade Risk

Introduction

Kessler syndrome is the central concept in the analysis of orbital debris. It links every other element of the debris problem, including mitigation, removal, tracking, policy, and economics, to a single underlying dynamic: the possibility that the debris population in some orbital region could become self-sustaining, growing through internal collisions even in the absence of further launches. The concept dates to a 1978 paper by Donald Kessler and Burton Cour-Palais of NASA's Johnson Space Center, and it has shaped the technical and policy response to debris ever since. Every modern simulation of the long-term debris environment, every argument for active debris removal, and every aggressive disposal requirement traces, directly or indirectly, to the analytical framework that Kessler and Cour-Palais established.

This chapter examines the original argument in detail, alongside the modern simulation literature that has refined and extended it, showing how closely the present orbital environment resembles the conditions Kessler described and the implications of cascade dynamics for the design of mitigation, removal, and policy responses. The treatment is necessarily technical in places, but the central conclusions can be stated in plain terms: the orbital debris problem is not a static inventory of risks but a dynamic system that responds to actions taken now; the most heavily affected altitude bands of low Earth orbit are at or near the threshold for self-sustaining cascade growth; and the response to the situation requires both preventing new debris generation and actively reducing the existing population. None of these conclusions is universally accepted in detail, but the consensus around their substance is broad enough to guide policy and engineering decisions across the main spacefaring states.

The Original Argument

The 1978 paper by Donald Kessler and Burton Cour-Palais was, in formal terms, an exercise in applying asteroid-belt evolution theory to the population of artificial Earth satellites. Kessler had spent the early 1970s studying the dynamics of the asteroid belt, applying mathematical techniques developed for analyzing granular collisional systems to the question of how a population of bodies on similar orbits evolves. The asteroid-belt analogy was not exact; asteroids orbit the Sun across enormous volumes, whereas artificial satellites are concentrated in a comparatively thin shell around Earth. However, the underlying mathematical structure of pairwise collision probabilities and fragment-generation dynamics translated reasonably well across the scale change.

By the mid-1970s, Kessler had begun to suspect that the same techniques, adjusted for scale and timescale, could be applied to the population of artificial satellites in low Earth orbit. The collaboration with Cour-Palais, an expert in hypervelocity impact at the Johnson Space Center, enabled Kessler to integrate population dynamics with realistic models of how collisions generate fragments. Cour-Palais had spent years studying the mechanics of high-velocity impact for spacecraft shielding applications. They brought to the collaboration a quantitative understanding of how impacts between two objects of given masses and velocities would produce a population of fragments spanning a range of sizes. The combination of population dynamics and fragmentation modeling produced a framework that, for the first time, could predict the long-term evolution of the orbital debris environment as a function of debris generation and removal rates.

The paper's central claim was simple. Consider a population of objects in similar orbits. Each object presents a cross-section that any other object can strike. The probability of a collision between any two specific objects over a given time is small. Still, the probability that a collision will occur in the population increases with the square of the number of objects, because the number of distinct pairs grows quadratically. When a collision occurs, it produces fragments, each of which is an object that can collide with the others. If the rate at which collisions produce new fragments exceeds the rate at which fragments are removed by atmospheric decay, the population grows. If the population grows, the collision rate rises. If the collision rate rises, the population grows further. The process is self-sustaining.

Kessler and Cour-Palais identified a critical density at which the cascade became unavoidable in each altitude band. Below the critical density, the system is stable: collisions occur, but atmospheric decay clears fragments faster than collisions create them. Above the critical density, the system is unstable: even if no further launches occurred, the population would continue to grow until natural processes eventually removed it on timescales of centuries or millennia. Critical density depends on altitude, with higher altitudes having lower thresholds because of weaker atmospheric drag. The 1978 paper estimated, using the launch rates and population data of the time, that some altitude bands could approach critical density within a century, and that the achieved density depended sensitively on the rates of fragmentation events, particularly the contributions from upper-stage explosions and anti-satellite tests.

The paper's projections were striking. At the launch rates of the late 1970s, and assuming no mitigation, the cascade process could become significant within decades rather than centuries if catastrophic fragmentation events continued. Kessler and Cour-Palais explicitly noted that the existing population of unpassivated upper stages, several of which had already exploded by the time the paper was written, represented a continuing source of fragments that could more quickly push the system toward instability than launch traffic alone would predict. The paper called for a systematic study of debris dynamics, measurement of the existing population, and consideration of mitigation measures to reduce the rate of fragment generation.

The paper was published in the Journal of Geophysical Research in June 1978 under the formal title "Collision Frequency of Artificial Satellites: The Creation of a Debris Belt" (citation 83(A6), pages 2637-2646). Kessler reportedly submitted the paper without explicit clearance from his NASA management, and its rapid, wide international circulation surprised the agency. Within a few years, NASA established the Orbital Debris Program Office at Johnson Space Center, with Kessler as its first head. The program produced the LEGEND model, Long-term Evolution and GENerational Debris, which, in its updated forms, remains the principal NASA tool for modeling the long-term evolution of the orbital debris environment. ESA's MASTER and DELTA models perform similar functions for the European context. Academic codes developed at universities and research institutes around the world provide additional independent validation. The basic structure of the original Kessler-Cour-Palais argument remains intact across these modern simulations, though the quantitative parameters and the treatment of specific physical effects have been refined.

The Mathematical Foundation

The mathematical structure underlying the cascade argument can be expressed in compact form. For a population N of objects with cross-section sigma (σ), mean relative velocity v, and spatial volume V, the rate of pairwise collisions is approximately $N^2\sigma v/2V$. Each collision produces an average of f new fragments above some size threshold. The fragments are removed at a rate proportional to the population, with a rate constant $1/\tau$, where τ is the mean lifetime of fragments at the relevant altitude. The net rate of change of the population is therefore approximately $fN^2\sigma v/2V$ minus N/τ.

Setting this expression equal to zero identifies the critical population at which generation and removal balance: N_critical = $2V/(f\sigma\tau v)$. Above this population, the right-hand side of the rate equation is positive, and the population grows. Below it, the population decays toward zero. The expression makes the dependence on altitude explicit through τ. At high altitudes, where τ is large, the critical population is low, so modest increases in density push the system into instability. At low altitudes, where τ is short, large populations can be sustained without triggering a cascade. The dependence on the fragment-generation factor f is also critical: an orbital environment in which collisions tend to produce many small fragments has a lower threshold than one in which collisions tend to produce fewer, larger fragments. The cross-section σ depends on the average size of objects in the population, with larger objects contributing more to the collision probability per object.

In plain terms, the formula says this: the safe threshold for any given altitude band depends on four factors. How large is the volume of space under consideration? How many fragments does each collision produce?

How long do those fragments survive before atmospheric drag removes them? And how large are the objects involved? High altitudes are the most dangerous not because they are more active, but because τ, the fragment lifetime, is enormous. A fragment at 800 kilometers altitude can persist for decades; one at 400 kilometers may be gone in a year. This difference in lifetime is why the cascade risk in the 800 to 1,000 kilometer band is far more serious than the cascade risk in the International Space Station's orbit, even though far more activity occurs near the station.

This simple model captures the essential physics but omits several refinements that are important for quantitative prediction. The fragment-generation function f depends on the energies of the colliding objects and on the mass distribution of the resulting fragments. The classical NASA breakup model, developed by Phillip Anz-Meador and colleagues, parameterizes the fragment-size distribution for a given collision in terms of the colliding masses and the relative velocity, with separate distributions for catastrophic collisions (in which both objects are completely fragmented) and non-catastrophic collisions (in which only one object is significantly damaged). The threshold for catastrophic fragmentation is conventionally defined by the impact energy-to-mass ratio: collisions with EMR above approximately 40 joules per gram are typically catastrophic.

The relative velocity v varies across the population, and collision energy scales as v^2. In low Earth orbit, the relative-velocity distribution peaks around 10 to 12 kilometers per second (6.2 to 7.5 miles per second), with notable contributions at both higher and lower velocities, depending on the inclination distribution of the colliding population. The distribution is asymmetric: most collisions occur between objects with similar inclinations at moderate relative velocities, but the small number of high-velocity collisions (between objects in retrograde orbits relative to each other) account for a disproportionate fraction of fragment generation because of the v^2 dependence of collision energy.

The spatial volume V is not uniform; debris is concentrated in thin shells at specific altitudes and inclinations, and the effective collision volume is much smaller than the geometric volume of the altitude band. Modern simulation codes explicitly account for this concentration by dividing the orbital volume into many small cells, each with separate population tracking. The cell-based approach reflects the reality that two objects in similar orbits at the same altitude but with different right ascensions of the ascending node may never come close to each other. In contrast, two objects on crossing orbits may repeatedly approach over many orbital periods. The full dynamics include both the long-term distribution of objects in altitude-inclination space and the short-term dynamics of conjunctions within that distribution.

Modern simulation codes explicitly incorporate these refinements, replacing the analytic expression with numerical integration over the actual population distribution. Time steps for propagating individual objects are typically on the order of days to weeks, with periodic recalculation of pairwise collision probabilities and statistical sampling of fragmentation events. Monte Carlo techniques are used to characterize uncertainty in long-term predictions, with hundreds or thousands of independent simulation runs producing a distribution of possible futures from which mean values and confidence intervals can be derived. Qualitative conclusions remain similar to those of the simple model: stable populations exist below a threshold, unstable populations above it, and the threshold is altitude-dependent. Quantitative refinements affect specific predictions about the rate of cascade growth and the identification of which altitude bands are at greatest risk. Still, they do not overturn the basic structure of the argument.

Modern Simulation Literature

Orbital debris simulation has been one of the more productive subfields of orbital debris research over the past four decades. NASA's LEGEND model, developed at the Johnson Space Center beginning in the early 2000s, traces its conceptual lineage directly to Kessler's original work and represents the most sophisticated American implementation of long-term debris environment modeling. ESA's MASTER (Meteoroid and Space Debris Terrestrial Environment Reference) and DELTA (Debris Environment Long-Term Analysis) models perform similar functions in the European context, using somewhat different methodological choices that provide useful independent validation. The Italian aerospace research center CIRA has developed the SDM (Space Debris Mitigation) model. Russian institutions have developed independent codes that incorporate a different population estimate based on Russian sensor data. Academic codes at the University of

Southampton, the Astronautical Engineering Institute at Beijing University of Aeronautics and Astronautics, the German Aerospace Center, and several U.S. universities supplement the agency models with independent analyses.

The convergence of these independent codes on similar conclusions provides some of the strongest evidence for the robustness of the cascade analysis. The codes differ in many implementation details, including the specific fragment-generation models used, the propagation methods for individual orbits, the treatment of perturbations, and the assumed launch traffic in future scenarios. Yet they consistently identify the most heavily affected altitude bands as supercritical, estimate cascade growth timescales as decades to centuries, and conclude that mitigation alone is insufficient to stabilize the long-term population. Disagreements among the codes typically concern growth rates in specific scenarios rather than the system's qualitative behavior.

A specific source of legitimate technical debate concerns how small fragments are treated in the simulations. The cataloged population of objects above approximately 10 centimeters (3.9 inches) in low Earth orbit is well measured and tracked. The medium-sized population, between 1 and 10 centimeters (0.39 to 3.94 inches), is estimated statistically from radar measurements of the broader environment but is not tracked individually. The small population, below 1 centimeter (0.39 inches), is estimated from impact measurements on the surfaces of returned spacecraft. Each of these populations contributes to cascade dynamics in distinct ways: large objects, when struck, produce thousands of fragments; small objects damage functional satellites without necessarily fragmenting them; medium objects bridge the two regimes. Different simulation codes weigh these contributions differently, and the overall predicted rate of cascade growth is highly sensitive to the weighting. As a result, quantitative predictions of the long-term population growth rate are highly uncertain even when the qualitative structure of the cascade is well established.

The simulation community has also begun explicitly incorporating the effects of large constellations. The deployment of Starlink, OneWeb, Amazon Leo, and the Chinese constellations has roughly doubled the number of active satellites in orbit over the past five years. These satellites are designed for end-of-life disposal and, if they perform as designed, will not meaningfully contribute to the long-term debris population. The principal sensitivity in the simulations is the failure rate: if a small fraction of constellation satellites fail to perform their disposal maneuvers, the cumulative contribution over decades of operations becomes increasingly important. ESA's recent MASTER updates, NASA's LEGEND scenarios, and several academic codes have run sensitivity analyses on constellation failure rates, and the consistent conclusion is that the long-term debris environment is sensitive to disposal performance at the percent level. A constellation that achieves 99 percent disposal compliance has a smaller long-term footprint than one that achieves 95 percent, and the difference accumulates over decades of operation.

How a Cascade Begins

It is useful to distinguish between two senses of the term "Kessler syndrome." In the strict, original sense, the syndrome refers to a self-sustaining process in which the in-situ collision rate between debris objects exceeds the rate at which natural decay removes them, leading to long-term population growth even in the absence of further launches. In the looser, more colloquial sense, the term describes any scenario in which the debris population grows fast enough to render an orbital region operationally unusable. The distinction matters because the strict sense involves a clear mathematical threshold. In contrast, the looser sense is a question of operational economics that depends on the level of risk operators are willing to accept.

A cascade, in the strict sense, is initiated when the rate of fragment production from collisions in a given altitude band exceeds the rate at which atmospheric drag removes fragments from that band. Initiation does not require a single dramatic event. It can occur through the gradual accumulation of objects until the collision rate, by then a relatively small statistical effect, becomes dominant. It can also be accelerated by a single dramatic event that adds many fragments at once. The Fengyun-1C anti-satellite test in 2007 added more than 3,500 trackable fragments to the 800- to 900-kilometer (497- to 559-mile) altitude band in a single moment. Analyses by NASA and ESA have repeatedly identified the test as the largest single contributor to the contemporary debris environment in low Earth orbit. The Iridium-Cosmos collision in 2009 added approximately 1,800 cataloged fragments to a similar altitude band. These two events, separated by two years, brought the affected altitudes notably closer to critical density than they otherwise would have been.

The dynamics of cascade initiation are subtle in ways that the simple analytic model does not fully capture. A single fragmentation event produces a debris cloud that initially occupies a narrow region of altitude-inclination space, a band of altitudes within a few hundred kilometers of the event altitude, within the parent object's orbital plane. Over weeks to months, the cloud spreads along the orbital plane as fragments disperse at slightly different velocities. Over months to years, the cloud spreads in altitude as fragments experience slightly different drag accelerations. Over years to decades, the cloud spreads in inclination as orbital planes slowly precess under J2 perturbations. The effective population available to participate in subsequent collisions, therefore, grows over time, even when the total fragment count is constant, as the cloud spreads to overlap with more of the orbital volume.

The implications for cascade dynamics are that an event creates not only an immediate increase in collision risk but also a long-term contribution that persists for decades. Fifteen years after the event, the Fengyun-1C cloud has spread to occupy a fraction of the orbital volume between roughly 600 and 1,200 kilometers (373 and 746 miles) at inclinations near 99 degrees, and it continues to spread as the orbital plane precesses. The Iridium-Cosmos cloud has similarly spread to occupy much of the volume between roughly 500 and 1,000 kilometers (311 and 621 miles) at inclinations near 86 degrees. The cumulative effect of these spread populations on long-term collision risk in low Earth orbit exceeds what would be predicted from the immediate fragment counts alone.

Modern simulations of the cascade process converge on several conclusions. First, the present-day environment in certain altitude bands of low Earth orbit, particularly between 800 and 1,000 kilometers (497 and 621 miles), is at or very near the threshold above which cascade growth occurs even without further launches. Second, the inclination structure matters; cascades develop most rapidly in the polar and sun-synchronous bands, where retrograde geometries produce high collision velocities. Third, cascade growth, even when triggered, is slow by human standards. Decades-long doubling times are typical in current models, with the most aggressive scenarios producing significant degradation over half a century rather than over a few years. Fourth, the cascade is not uniformly distributed: even when the overall low Earth orbit becomes supercritical, growth is concentrated in specific altitude-inclination bands, leaving other regions relatively unaffected.

How Close Are We?

The question of how close the present environment is to a runaway cascade is among the most actively studied topics in debris research. The answer, with appropriate caveats, is that the most heavily affected altitude bands in low Earth orbit are at or near the threshold, but that the rate of cascade growth, if it has begun, is slow. Several independent lines of evidence support this assessment.

First, population statistics show steady, though not exponential, growth in the cataloged debris population over the past two decades, after accounting for discrete contributions from growing fragmentation events. The increase caused by in-situ collisions, distinct from launches and large fragmentation events, remains small but measurable. ESA's analyses estimate that the contribution of unintended collisions to the population is, at present, of the same order as that of launches and explosions, and may overtake them in the coming decades if mitigation does not improve. The transition from a launch-dominated regime to a collision-dominated regime is a qualitative milestone identified in the cascade literature. When most new debris is produced by collisions among existing debris rather than by new launches, the system has crossed a threshold that distinguishes launch-driven environment growth from cascade-driven environment growth.

Second, the rate of operational conjunctions, defined as close approaches between cataloged objects that require avoidance maneuvers, has grown dramatically over the past decade, driven by increases in both the populations of tracked objects and active satellites. The Combined Space Operations Center, which provides conjunction data services to operators worldwide, processes hundreds of thousands of close-approach warnings per day, and hundreds of these result in actual avoidance maneuvers each year by satellites in the commercial constellations alone. SpaceX has reported that its Starlink constellation performed tens of thousands of avoidance maneuvers in 2023 and 2024, with the rate climbing steadily as both the constellation and the surrounding debris environment have grown. Each avoidance maneuver is, in a sense, a near miss; the underlying pattern of close approaches reflects the density of the environment more honestly than population statistics alone.

Third, the intuitive sense that debris generation is outpacing removal is increasingly supported by quantitative analysis. ESA's 2025 Space Environment Report notes that more than 3,000 trackable fragments were added to the catalog in 2024 alone, while approximately 2,031 objects reentered. The balance is delicate, and small changes in either generation or removal rates would tip it. The encouraging fact is that mitigation discipline has improved over the same period: compliance with disposal guidelines has risen, commercial operators have invested in autonomous collision avoidance, and the regulatory environment has tightened. The discouraging fact is that adverse events, including anti-satellite tests, propulsion failures during disposal maneuvers, and loss-of-control scenarios on aging satellites, could produce step changes that overwhelm the improvements.

Fourth, modeling studies that project the current environment forward under realistic launch traffic and mitigation compliance show population growth in most scenarios. NASA's standard projection assumes mitigation guidelines are followed at current compliance rates and that no major fragmentation events occur. In this scenario, the cataloged population in low Earth orbit grows by approximately 30 percent over the next half-century, with growth concentrated in the most heavily affected altitude bands. ESA's projections are qualitatively similar but differ somewhat in quantitative details. Aggressive mitigation scenarios, in which compliance rises to 99 percent, no anti-satellite tests occur, and several active removal missions are conducted, stabilize the population near current levels. Pessimistic scenarios, in which compliance falls, anti-satellite testing continues, and removal capability fails to develop, produce substantial growth and approach the conditions of a recognizable cascade within decades.

The Stability Question

A specific technical question that has emerged from the modern simulation literature is whether the current low Earth orbit environment is stable, marginally stable, or already supercritical in the strict cascade sense. Different models yield somewhat different answers, in part because they make different assumptions about future launch traffic, mitigation compliance, and removal capacity. A frequently cited consensus, articulated most clearly in NASA studies of the late 2010s and reaffirmed in later ESA work, holds that the environment in the most heavily affected altitude bands, around 800 to 900 kilometers (497 to 559 miles), is supercritical even without further launches. That is, cascade growth would continue indefinitely if all human activity in low Earth orbit ceased today. The growth would be slow, on the order of a percent per year, but positive.

If this consensus is correct, mitigation alone is insufficient. Preventing the addition of new debris is necessary but not sufficient; the existing population in the supercritical bands must be reduced. This is the technical case for active debris removal, the class of missions discussed in detail in Chapter 11. The case is not that removal alone solves the problem, but that without removal, the problem grows even when prevention succeeds. The supercritical conclusion shapes the priorities of the international debris community in important ways. For example, it makes removal capability a strategic priority rather than a desirable supplement. It identifies specific altitude-inclination bands as the regions where removal efforts would yield the greatest benefit.

The supercritical conclusion is not universally accepted, and several careful analysts have argued that it depends on assumptions about the rate of small-fragment generation that may be overly optimistic. The argument runs as follows: most simulations assume that catastrophic collisions produce fragments according to the NASA breakup model, which was calibrated against ground tests of relatively low-energy impacts and may not accurately reproduce the behavior of high-energy impacts in space. If actual fragmentation produces fewer small fragments than the model predicts, the cascade contribution from each collision is smaller, and the threshold for supercritical behavior is higher than the simulations suggest. This argument has been contested by other analysts, who note that subsequent in-orbit fragmentation events have produced fragment counts consistent with the model predictions and that the data, if anything, suggest more fragmentation rather than less.

The honest summary is that the question is not closed; the most authoritative analyses point to a marginally supercritical or supercritical environment in the most affected bands; and the practical policy implications are robust to most remaining uncertainty: mitigation alone is inadequate, removal capability needs to be developed, and the conditions that would tip the environment toward more rapid growth need to be avoided. Whether the present environment is exactly at the threshold, slightly above it, or slightly below it matters less for policy than the consensus that it is close enough that adverse events could push it decisively into instability.

Cascade Geometry and the Inclination Problem

The cascade dynamics are not uniformly distributed across the orbital volume. In addition to the altitude dependence discussed above, the cascade is highly sensitive to inclination. Two objects in nearly identical orbital planes can occupy similar altitudes for years without ever coming close to each other because their phase relationship, the relative position along their orbits, keeps them apart. Two objects in crossing orbital planes, by contrast, approach each other twice per orbit at the points where their orbital planes intersect. The relative velocity at these encounters depends on the angle between the planes: small inclination differences produce small relative velocities, whereas large inclination differences produce velocities approaching twice the orbital velocity.

The historical concentration of operational missions in specific inclination bands has produced corresponding concentrations of debris. Sun-synchronous orbits near 98 degrees host the bulk of Earth observation, weather, and reconnaissance missions and consequently contain much of the long-lived debris from those missions. Other reconnaissance platforms and similar missions use polar orbits at 80-90 degrees. The intersections of these orbital planes occur at high latitudes and at high relative velocities, producing the geometric conditions most conducive to cascade growth. Modeling studies consistently identify the polar and sun-synchronous bands of low Earth orbit between 700 and 1,000 kilometers (435 and 621 miles) as the regions where the cascade would first develop. The events of recent decades, the Fengyun-1C and Iridium-Cosmos events, both of which occurred in these bands, have brought these regions closer to that threshold.

The inclination geometry has implications for the design of mitigation and removal strategies. Active debris removal missions targeting objects in sun-synchronous orbits can operate efficiently because the target population is concentrated in a narrow inclination band accessible from a single launch trajectory. Multi-target missions that capture and dispose of several objects on a single mission can be designed to chain through targets in similar orbital planes, minimizing plane changes between captures. The ESA ClearSpace-1 mission, retargeted in 2024 to the PROBA-1 satellite, exemplifies this approach: the target was selected in part for its accessibility from a single launch trajectory, and the mission profile keeps the removal vehicle within a manageable plane-change budget throughout its operations.

The inclination geometry also has implications for the deployment of mega-constellations. Most constellation operators have chosen inclinations that minimize plane-change costs from their launch sites and provide adequate latitude coverage for their service requirements. Starlink, launched from Cape Canaveral, uses inclinations between 53 and 70 degrees, achievable from Florida without prohibitive plane changes. OneWeb, launched from Kazakhstan, uses inclinations near 87.4 degrees, consistent with launch from Baikonur. The geographical clustering of operational constellation altitudes around specific inclination bands creates new conjunction concerns within those bands, and operators have invested heavily in coordination protocols to manage the risks. The next generation of constellations will need to address how their orbital planes interact not only with one another but also with the long-lived debris populations from earlier eras.

What a Severe Cascade Would Look Like

The most extreme scenarios in the cascade literature, sometimes called the ablation cascade or debris belt formation scenarios, describe outcomes in which one or more altitude bands of low Earth orbit become so densely populated with fragments that operations in those bands become impractical. The Iridium-Cosmos collision in 2009 produced a fragment cloud that took weeks to spread along the orbital plane and years to disperse across the affected altitude band fully. A series of such collisions, each contributing to and being contributed to by the others, could, in principle, produce a population dense enough to make conjunction screening unmanageable, force continuous avoidance maneuvers, and eventually render operating in the band prohibitively expensive. The affected band would not be physically impassable; large fractions of the volume would still be empty. But the operational risk would exceed the value provided by missions in that band, and, by economic logic, the band would be abandoned.

A scenario from the simulation literature describes the cascade as follows. An initial fragmentation event, a collision between a derelict satellite and a fragment from an earlier event, adds several thousand new fragments to a band already near the cascade threshold. The new population pushes the band into supercritical territory. Over the next decade, the elevated population will produce a small but measurable increase in the rate of additional collisions. Each new collision adds a few thousand fragments, further increasing the population and the collision rate. The population's doubling time is initially slow, perhaps 50 years, but accelerates as the population grows because the collision rate scales with the square of the population. By a century after the initiating event, the affected band may contain ten times the original debris population, and the rate of secondary collisions may have grown to one or more per year. By this point, the operational consequences would be severe: constant avoidance maneuvers would be required, and the cumulative propellant cost would make long-duration missions impractical.

The timescale on which this scenario unfolds, even in the most aggressive models, is on the order of decades to centuries. There is no scenario in the current literature in which low Earth orbit becomes unusable within a year, or even within a decade, absent a deliberate large-scale fragmentation event. The cascade, if it occurs, is slow. This slowness is both a consolation and a trap. It is a consolation because it gives policy responses time to develop and deploy. It is a trap because the slow accumulation of consequences, spread over decades, creates persistent incentives for individual actors to defer costs to the future and free-ride on others' actions. The cascade is, in this sense, a textbook commons problem, with all the political and economic challenges that the commons presents.

A useful comparison is to climate change. Both problems involve the cumulative effect of distributed activity on a shared environmental resource. Both unfold on timescales longer than human attention spans yet short enough to matter for the lifetimes of present infrastructure and institutions. Both require cooperation among states with divergent interests and capabilities. Both face powerful incentives for individual actors to externalize costs. The orbital debris problem is, fortunately, more tractable than climate change in several respects: the affected community is smaller, the technical interventions are better defined, and the relevant response timescales are longer. But the structural similarity is real and suggests that the institutional responses developed for climate change may offer lessons for debris policy. Both fields, for example, have struggled with how to allocate responsibility for cumulative damage among actors who contributed at different times and to varying degrees, and both have developed concepts of common but differentiated responsibility that may translate.

Operational Reality Today

Operators of satellites in low Earth orbit face the cascade question daily. SpaceX has reported that its Starlink constellation has performed tens of thousands of collision-avoidance maneuvers over the past several years; the rate has steadily climbed as both the constellation and the surrounding debris environment have grown. The International Space Station has performed dozens of debris-avoidance maneuvers over its operational life, including several prompted by fragments from anti-satellite tests in recent decades. The November 2021 fragmentation of Cosmos 1408 forced the ISS crew to take shelter in their return spacecraft as the fragment cloud passed nearby. The event was widely reported and brought the operational reality of the debris environment to mainstream public awareness.

Beneath these visible events, the underlying conjunction screening infrastructure has expanded dramatically. The U.S. Space Surveillance Network now provides conjunction data services to operators worldwide through the Eighteenth Space Defense Squadron and the Combined Space Operations Center. ESA operates an independent space surveillance and tracking program through its Space Safety Programme. Commercial space situational awareness providers, including LeoLabs, NorthStar Earth and Space, ExoAnalytic Solutions, Slingshot Aerospace, and others, supplement government tracking with phased-array radars and optical telescopes purpose-built for debris observation. The combined infrastructure now processes far more data than any single national service did even a decade ago. Still, it remains stretched, and several recent incidents have illustrated the consequences of late warnings or inadequate uncertainty information.

The operational consequences of cascade-level density are increasingly evident in the day-to-day operations of leading satellite operators. Starlink's average satellite performs more than one avoidance maneuver per year, and the rate has steadily increased as the constellation and its surrounding environment have expanded. Each maneuver consumes propellant that would otherwise extend operational life. Each maneuver interrupts service to ground users. Each maneuver requires an operator to decide whether the probability of conjunction justifies the operational disruption. The cumulative cost across the constellation is material, though it remains small relative to the value of the services provided. As the debris environment grows, the cost-to-value ratio shifts unfavorably; the point at which it becomes prohibitive depends on the trajectory of debris growth and the unit economics of the constellation operators.

Mathematics of Conjunction Risk

The operational management of cascade risk depends on probability calculations that quantify the likelihood of collision for any given close approach. The standard expression for the probability of collision (Pc) is computed from the predicted positions of the two objects, the uncertainty distributions of those positions, and the combined cross-sectional area of the two objects. The calculation is performed for every cataloged conjunction within a screening volume; objects within large commercial constellations may undergo thousands of such calculations per day. When the computed probability of collision exceeds an operator-defined threshold, typically 1 in 10,000 to 1 in 100,000 for crewed and high-value spacecraft, with looser thresholds for some commercial operations, the operator considers a maneuver.

To give those numbers a human scale: a 1 in 10,000 collision probability means that if the same conjunction were repeated ten thousand times under identical conditions, you would expect one collision. Across a constellation of 10,000 satellites, each performing multiple conjunctions per day, even a 1-in-100,000 threshold still translates into a small number of expected collisions per year if it is the only protection. This is why conjunction probability thresholds are not the only defense. Shielding, redundancy, and the physical miss distance all matter. The probability threshold is the trigger for a maneuver decision, not a guarantee of safety on its own.

The calculation proceeds as follows. The predicted positions of the two objects at closest approach are computed by propagating their orbital elements forward from the most recent observations. The position uncertainties at closest approach are computed similarly, accounting for uncertainty in the orbital elements and the propagation interval. The relative position vector and its uncertainty distribution are obtained by subtracting the two predicted positions and combining their uncertainty distributions. The probability of collision is the integral of the relative-position probability density function over the volume where the two objects' physical extents would overlap. For typical satellite-on-satellite conjunctions, the integral is approximately the relative-position probability density at zero relative position, multiplied by the combined cross-sectional area of the two objects.

Several technical issues complicate the calculation in practice. Position uncertainties are not symmetric: they are typically larger along the track than in the cross-track or radial directions because along-track position is most sensitive to drag and ephemeris errors. The probability of collision can therefore be underestimated when uncertainties are large, even though larger uncertainties might intuitively seem to reduce the apparent collision risk. This phenomenon is sometimes called the dilution problem, and it has produced several near-collisions in the historical record, including the 2009 Iridium-Cosmos event. The remedy is to require that uncertainty be small, not merely that the nominal miss distance be large, before deciding not to maneuver. Modern conjunction analysis tools incorporate these considerations, but the underlying physics ensures that the conjunction-screening process is inherently uncertain, and the operational community has had to accept and design around an irreducible level of risk.

A related complication is how the conjunction prediction evolves as the time of closest approach approaches. A conjunction predicted seven days in advance carries large uncertainties because the propagation interval is long. As new observations become available, the predicted positions and uncertainties are refined. The probability of collision may rise or fall as new data comes in. Operators must decide when to commit to a maneuver: a decision made too early may prove unnecessary if later data resolves the conjunction. At the same time, a decision made too late may not leave enough time to execute the maneuver reliably. The optimal timing depends on the spacecraft's maneuver capability, the mission's operational constraints, and the rate at which new tracking data becomes available. Modern operations centers run conjunction calculations continuously and update maneuver planning as new data becomes available.

Maneuver decisions involve tradeoffs that extend beyond the probability of collision. A maneuver consumes propellant, reducing operational lifetime and causing lost revenue. A maneuver may interrupt operations: an Earth observation satellite cannot collect imagery during the maneuver, a communications satellite may need to redirect traffic, and a scientific mission may need to revise its observation schedule. These costs are concrete and quantifiable. By contrast, the cost of not maneuvering is the small probability of catastrophic loss, which translates statistically into an expected cost. The ratio between these costs, together with the operator's risk tolerance and the precision of the conjunction analysis, determines the operational thresholds for maneuver decisions.

The Role of Active Debris Removal

The cascade analysis provides the technical case for active debris removal, which the policy community has increasingly accepted. If the most heavily affected altitude bands of low Earth orbit are at or above the supercritical threshold, mitigation alone, however perfectly implemented, cannot prevent long-term population growth. The existing population in those bands must be reduced. The arithmetic of removal is straightforward in principle. Each large object removed eliminates both the risk of a direct collision and the potential for thousands of additional fragments if struck. Modeling studies have repeatedly shown that the marginal benefit of removing a single large object can exceed the marginal cost of preventing several additional fragmentation events, and that the benefits scale favorably with the number of objects removed.

Cascade analysis identifies which objects should be prioritized for removal. Simulations then determine which specific objects, if removed, would yield the greatest reduction in long-term collision probability across the affected altitude bands. The rankings depend on the object's mass (which determines the number of fragments it would produce if struck), its altitude (which determines how long the resulting fragments would persist), and the local density of objects with which it might collide (which determines its probability of being struck). The top of the resulting priority list is dominated by large derelict objects in the polar and sun-synchronous bands at altitudes between 700 and 1,000 kilometers (435 and 621 miles), including Soviet-era SL-16 rocket bodies, defunct ENVISAT and similar large Earth observation satellites, and specific accumulations of debris from past fragmentation events.

The economics of removal, examined in detail in Chapter 13, are challenging. Each removal mission costs tens to hundreds of millions of dollars and removes one or a few objects. The total population requiring removal to stabilize the most affected bands is on the order of dozens of objects, a manageable number in principle but a substantial commitment of resources in practice. Whether the necessary removal capability will be developed at the scale and pace suggested by the cascade analysis depends on a combination of technical maturation, regulatory pressure, and the willingness of national governments to fund missions whose direct beneficiary is the entire international community rather than the funding state alone. The ClearSpace-1 and PRELUDE missions mark the first wave of operational demonstrations; the question is whether they will be followed by a sustained capability or whether removal will remain a demonstration activity.

The Role of Mitigation

Mitigation, the prevention of new debris generation, remains the foundation of any plausible response to the cascade dynamic. Even a fully developed removal capability would be overwhelmed if mitigation failed and new debris were generated faster than it could be cleared. The relative leverage of mitigation versus removal varies by scenario: in optimistic mitigation scenarios with no notable fragmentation events, mitigation alone may stabilize the population without large-scale removal; in pessimistic mitigation scenarios with ongoing anti-satellite tests and explosions, even ambitious removal programs cannot keep pace.

The principal mitigation interventions are end-of-life disposal, passivation of upper stages, and prevention of operational debris release and fragmentation events. End-of-life disposal removes the satellite from the debris-vulnerable region within a defined timeframe after mission completion, either through atmospheric reentry (for satellites at altitudes where this is achievable through propulsion or natural decay) or by transferring to a graveyard region (for satellites at altitudes where reentry is not practical). Passivation removes the explosion risk from upper stages by venting residual propellants after spacecraft separation. Preventing the release of operational debris eliminates the deliberate release of mission-related items during normal operations. Preventing fragmentation events addresses the strategic and operational decisions that lead to anti-satellite tests, deliberate breakups, and collisions.

The first three categories, disposal, passivation, and operational debris control, are well established and addressed by standard design practices in modern spacecraft. The remaining category, prevention of fragmentation events, is the principal ongoing source of new debris and the most resistant to technical solutions. Anti-satellite tests are matters of strategic policy, not engineering. Accidental collisions can be reduced through better tracking and conjunction screening, but cannot be eliminated if significant populations of objects share orbital regions. The mitigation discipline, therefore, continues to evolve, with the technical community focused on improving the effectiveness of established interventions and the policy community focused on constraining events that technical interventions cannot address.

Why the Cascade Is the Central Concept

Kessler syndrome occupies a prominent place in any treatment of orbital debris, not because the catastrophic scenario is imminent. It is not. Rather, the cascade concept connects every other element of the debris problem to a single underlying dynamic. Mitigation matters because every fragment kept from entering the population reduces the cascade's contribution. Removal matters because the supercritical character of certain bands means the population must be actively reduced rather than merely held constant. Tracking matters because conjunction avoidance is the operational defense against the increased collision rate that a higher population produces. Policy matters because the cascade's commons character means individual actors have inadequate incentives to bear the costs of preventing it. Economics matter because the cost of operating in a degraded environment is the mechanism by which cascade consequences are transmitted to operators and, ultimately, to users of satellite services.

Cascade is not the only concept in debris literature, and its importance can be overstated. Many practical debris problems, including collision avoidance, end-of-life disposal, and the routine management of conjunction warnings, would exist whether or not cascade dynamics were operative. But the cascade is why no single intervention can solve the problem and why it cannot be deferred indefinitely. It transforms the debris environment from a static inventory of risks into a dynamic system that responds to actions taken now and, over the coming decades, will either be brought under control or further degrade.

The cascade is also the conceptual bridge between the technical and policy communities that work on debris. The technical community can simulate cascade dynamics, identify the conditions under which growth occurs, and quantify the marginal effects of different interventions. The policy community can design instruments, licensing rules, international guidelines, and treaty provisions that align individual actors' incentives with the collective interest in preventing cascades. Neither community alone can solve the problem; technical interventions require policy support to be implemented at scale, and policy interventions require technical analysis to identify which are most effective. The cascade concept provides the common language for both communities.

Future Trajectories and Open Questions

The trajectory of the orbital debris environment over the coming decades hinges on a small number of decisions and developments. The rate of new debris generation depends on whether anti-satellite testing resumes at high altitudes, whether passivation discipline is maintained as new launch vehicles enter service, and whether constellation disposal performance meets the high-percentage thresholds assumed in the simulations. The rate of debris removal depends on whether the demonstration missions of the late 2020s mature into a sustained removal capability, whether the international community develops financing mechanisms adequate to support a meaningful removal program, and whether the technical maturation of removal capabilities continues to lower the cost per object. The rate of growth in the active satellite population depends on whether the constellation deployments now planned proceed as scheduled, whether new constellation designs incorporate aggressive disposal requirements, and whether the regulatory environment continues to tighten.

Several open questions are particularly consequential. The first concerns whether the international norm against destructive anti-satellite testing, evidenced by the 2022 American commitment and endorsed by more than thirty other states, will endure over the coming decade. Russia and China, the two leading spacefaring states that have not endorsed the norm, retain anti-satellite capabilities and have demonstrated a willingness to use them. A single high-altitude test by a non-endorsing state could significantly alter the cascade trajectory of the affected altitude bands. The second concerns whether the largest commercial constellation operators will meet the disposal performance requirements of their licenses. Even a small fraction of failed disposals, multiplied across thousands of satellites and decades of operations, would make a significant contribution to the debris field. The third concerns whether removal capability will scale to the volume the cascade analysis suggests is needed. Single-target demonstration missions are necessary but not sufficient; sustained removal at a rate of several large objects per year would require an operational and financial model different from that established in the demonstration era.

The fourth question, perhaps the most fundamental, is whether the international system can effectively manage a commons problem as large and long-term as the orbital debris environment. The technical case for action is well established. The policy instruments that would translate the technical case into effective action, including licensing rules, international guidelines, treaty provisions, and financing mechanisms, exist in various forms and with varying degrees of effectiveness. The remaining gap is political: the alignment of national interests is sufficient to sustain action over decades. The history of similar commons problems, such as climate change, ocean fisheries, and ozone depletion, provides mixed precedents. Some have been managed successfully through international cooperation; others have not. The orbital debris problem is, in many respects, more tractable than these analogs. Still, tractability does not guarantee action, and the question of whether the international community will act decisively enough remains open.

Figure 4.1. Logistic-Style Growth Curves Illustrating the Cascade Dynamic in Supercritical Altitude Bands

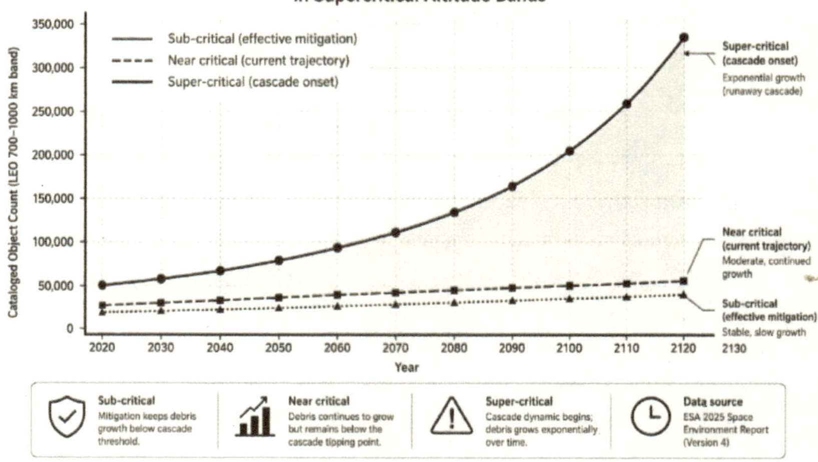

Figure 4.2. Cataloged Fragments from Major Debris-Generating Events, 1985–2022

(Includes cataloged fragments ≥10 cm in size)

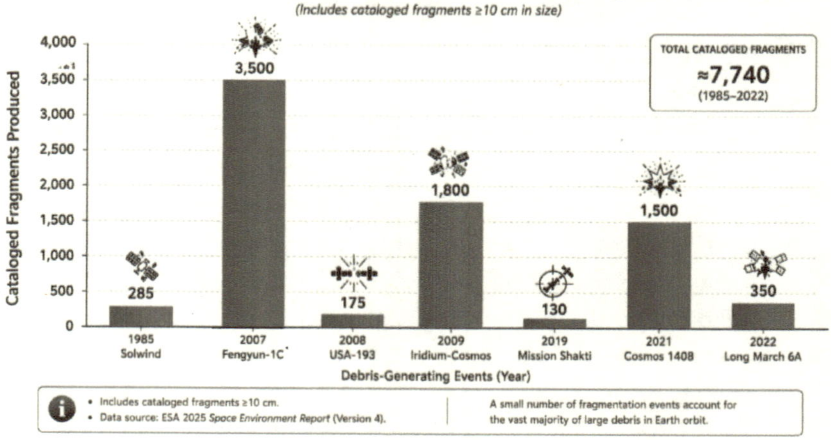

Altitude Band	Spatial Density (mid-2026)	Critical Density (Approximate)	Status
400–500 km 249-311 Miles (ISS regime)	Moderate	Well above critical due to atmospheric drag	Sub-critical (drag-cleared)
700–900 km 435-559 Miles (most affected)	Highest in LEO	Estimated near or at critical	Near critical/debatable supercritical
Above: 1,200 km 746 Miles	Lower per volume	High due to long persistence	Variable; some bands supercritical

Table 4.1. Critical Parameters for Cascade Analysis at Three Representative Altitudes

Chapter Summary

The Kessler syndrome describes the conditions under which the orbital debris population in a given altitude band can become self-sustaining, growing through internal collisions even without further launches. The original Kessler-Cour-Palais argument of 1978 has been refined by decades of subsequent simulation work, but its essential conclusions remain. The most heavily affected bands of low Earth orbit, particularly between 800 and 1,000 kilometers (497 and 621 miles), are at or near the critical density above which cascade growth occurs. If it has begun, the growth is slow on human timescales. Still, it is positive and transforms the debris problem from a static inventory of risks into a dynamic system requiring active management. The mathematical structure of the cascade, the modern simulation literature that has refined it, and the operational realities of conjunction screening together show that mitigation alone is insufficient and that active removal capability must be developed to stabilize the most affected regions. The cascade is the reason the orbital debris problem is structural rather than incidental, and it is the framework within which every other element of the debris response, including mitigation, removal, tracking, policy, and economics, must be understood. The chapters that follow examine the specific events that have brought the environment to its current state, the policies and technologies that aim to manage the dynamic, and the longer-term questions of governance and sustainability that the cascade raises.

Chapter 5. Major Fragmentation Events: Case Studies

Introduction

The orbital debris population has been shaped more than by any other factor through a relatively small number of discrete fragmentation events. Two-thirds of all cataloged objects in low Earth orbit can be traced to fewer than fifty events. Understanding these events, their causes and consequences, and the policy responses they precipitated is essential to understanding the current state of the orbital environment and the prospects for managing it in the future. The case studies that follow are not merely historical; they form the empirical foundation on which the modern technical understanding of fragmentation rests and have shaped the policy frameworks within which contemporary space activity operates.

This chapter examines the most significant events of the satellite era as case studies. The selection focuses on events that have most shaped the debris environment quantitatively, those that added thousands of cataloged fragments to the population; events that have most shaped the policy environment, those that demonstrated the consequences of particular categories of decisions; and events that have most shaped the operational practices of the satellite community, those that revealed gaps in tracking, conjunction screening, or mitigation procedures. The case studies are presented in approximate chronological order, with each treated at sufficient depth to draw out its specific lessons. Several smaller but instructive events are also included, both to illustrate categories of debris generation not captured by the largest events and to show how the cumulative pattern of smaller incidents contributes to the overall debris environment.

The Soviet Co-Orbital Anti-satellite Tests, 1968-1982

The history of anti-satellite testing begins not in the 1980s with the better-known American and Chinese events, but in the 1960s with the Soviet Istrebitel Sputnikov ("Satellite Destroyer") program. Between 1968 and 1982, the Soviet Union conducted approximately twenty co-orbital anti-satellite intercept tests, placing an interceptor satellite in an orbit similar to that of a target satellite, then maneuvering it into close proximity to detonate a fragmentation warhead. The tests were conducted at altitudes ranging from approximately 500 to 1,000 kilometers (311 to 621 miles), within inclination bands chosen for the strategic importance of the targets the program was intended to threaten.

The Istrebitel Sputnikov tests were classified at the time and remain incompletely documented. Western intelligence assessments compiled during and after the Cold War have identified specific test events from observed fragmentation signatures detected by the U.S. Space Surveillance Network. However, the Soviet program records have not been fully released. The program's cumulative debris contribution has been estimated at several hundred cataloged fragments, with more remaining below the tracking threshold. Most fragments have decayed over the decades since the tests, but a significant population persists at altitudes near 1,000 kilometers (621 miles), where decay is slow.

The surviving fragments from the Istrebitel Sputnikov program are particularly difficult to attribute with certainty because the Soviet Union did not publicly acknowledge specific test events during the Cold War, and the early American tracking infrastructure lacked the precision to catalog each event in real time fully. Some of what appears in the current debris catalog as "unknown origin" fragmentation from the 1970s is almost certainly material from these tests. The best current estimate, based on a retrospective analysis of fragmentation signatures in the tracking record, is that the program contributed between 500 and 900 long-lived fragments at altitudes above 900 kilometers, where natural decay operates on century timescales.

The strategic context of the Istrebitel Sputnikov program is worth noting because it established patterns of behavior that have shaped subsequent anti-satellite programs. The Soviet Union conducted the tests as part of a broader strategic competition with the United States, in which the use of orbit for reconnaissance, navigation, and early warning had become fundamental to both nations' nuclear postures. The capability to threaten American satellites was understood as a deterrent against American anti-satellite capability, even though the operational utility of co-orbital intercepts in actual conflict was always limited by the time needed to maneuver an interceptor to its target. The tests were primarily political demonstrations rather than operational rehearsals, and the consequences of the debris were deemed acceptable because the strategic benefit of demonstrating capability was paramount. This pattern, in which strategic demonstration overrides environmental consequences, has recurred in every subsequent anti-satellite testing program.

The 1985 ASM-135 Test Against Solwind P78-1

On September 13, 1985, the United States conducted the first successful direct-ascent anti-satellite intercept in its history. An ASM-135 missile launched from an F-15 destroyed the Solwind P78-1 solar physics satellite at an altitude of approximately 555 kilometers (345 miles). The intercept was the culmination of a development program that began in the 1970s as part of the broader strategic competition with the Soviet Union, which had been conducting co-orbital anti-satellite tests since 1968.

The technical character of the ASM-135 differed substantially from the Soviet co-orbital approach. Rather than placing an interceptor in orbit and maneuvering it to the target, the ASM-135 used a kinetic kill vehicle launched directly from a high-altitude aircraft. The kill vehicle, accelerated to the necessary velocity by the missile's solid rocket motor, intercepted the target on a single-ascent trajectory. The approach was operationally simpler than the Soviet approach; no interceptor satellite needed to be kept in orbit, but it produced equivalent fragmentation consequences when applied against a target.

The Solwind P78-1 satellite, launched in 1979, had been operating well beyond its design lifetime by 1985 and was nearing the end of its useful life. However, the choice of target was not entirely opportunistic; Solwind had been an important platform for solar physics, and the scientific community had not been consulted before the decision to use it as an anti-satellite target. The decision was made within the U.S. Air Force test program, with limited interagency coordination, and drew intense criticism from the scientific community and international observers, who viewed the destruction of an operational scientific satellite for testing purposes as needlessly provocative.

The event produced approximately 285 cataloged fragments. Because the intercept altitude was relatively low, atmospheric drag cleared most fragments within a decade. The last cataloged fragment from the test was reentered in 2002, seventeen years after the event. During their lifetimes, the fragments did not cause known damage to other satellites. Still, they were a serious concern at the time and contributed to the international policy debates that eventually led to the IADC mitigation guidelines of 2002.

The lessons of the ASM-135 test were several. First, anti-satellite testing produced large numbers of cataloged fragments, even at altitudes where atmospheric drag would eventually clear them. The 285 fragments cataloged from the single Solwind test represented a substantial addition to the mid-1980s background debris population. They proved unambiguously that direct-ascent kinetic intercepts produced fragmentation effects comparable to those of the Soviet co-orbital approach. Second, the choice of altitude mattered profoundly: a similar test at a higher altitude would have left long-lived debris with consequences extending across decades. Third, the test's political fallout, both domestically and internationally, contributed to later American decisions to avoid further direct-ascent destructive testing for more than two decades. The pattern of restraint did not survive the changed strategic environment of the 2000s, but it set a precedent that would later be cited in the post-2007 debates over Chinese anti-satellite testing.

The U.S. Congress passed legislation after the ASM-135 test that limited further American anti-satellite testing, and the program was effectively ended within a few years. The constraint was not a binding international commitment, and no equivalent constraints applied to other states' programs. Still, American restraint was visible internationally and affected the strategic calculations of other nations considering similar capabilities. The pattern of one state's restraint influencing others' calculations would prove inconsistent in the following decades, but the precedent had been established.

The 1996 Cerise-Ariane Collision

On July 24, 1996, the French military reconnaissance satellite Cerise was struck by a fragment from the upper stage of an Ariane 1 rocket that had exploded several years earlier. The impact occurred at an altitude of approximately 660 kilometers (410 miles) and severed Cerise's gravity-gradient stabilization boom. The satellite survived the impact and was eventually returned to operational service via ground commands to compensate for the loss of the stabilization boom. Still, the collision was the first confirmed accidental hypervelocity collision between two cataloged objects in orbit.

The Cerise event went largely unnoticed at the time but is noteworthy in retrospect for several reasons. It was empirically demonstrated that the cataloged debris population posed real operational risks to active satellites. This conclusion had been theoretically obvious but had not previously been proven by actual damage. It showed that even small fragments could cause serious damage to functional satellites, with the boom-severing incident illustrating that hypervelocity impacts can disable specific subsystems even when the satellite as a whole survives. It also established that the operational management of debris risk needed to extend beyond catastrophic collision avoidance to include consideration of damage to specific external structures.

The fragment that struck Cerise was traced to the November 1986 breakup of an Ariane H-10 upper stage, ten years before the Cerise event. The H-10 fragmentation was itself a pivotal event in the historical record, producing several hundred cataloged fragments that persisted in low Earth orbit for years. The connection between the 1986 fragmentation and the 1996 collision illustrated the long-term consequences of fragmentation events: a fragment generated in one decade could damage a satellite operating a decade or more later, in some cases, a satellite operated by a different state or commercial operator. The principle of intergenerational debris consequences, that fragmentation events affect not only the operators present at the time of the event but also the operators who would have used the affected orbital region in subsequent years, was implicit in the Kessler analysis of 1978 but became operationally visible through the Cerise event.

The 2007 Briz-M Upper Stage Explosion

On February 19, 2007, a Russian Briz-M rocket upper stage exploded at altitudes spanning a wide range of geostationary transfer orbits. The Briz-M is a third stage used with Proton launch vehicles for missions requiring high-energy trajectories, including geostationary deployments and translunar missions. The stage had been launched in February 2006 and had been in orbit for about a year when its propellant systems failed catastrophically. The event produced approximately 1,000 cataloged fragments distributed across an extensive volume of orbital space.

The Briz-M explosion was particularly consequential because of the parent stage's orbital geometry. Briz-M stages used for geostationary transfer follow highly elliptical orbits with perigees in low Earth orbit and apogees near geostationary altitude. The fragments from the explosion showed a similarly broad range of orbital geometries, with debris spanning altitudes from a few hundred kilometers to nearly 36,000 kilometers (22,369 miles). This broad distribution complicated the cataloging and tracking of the fragments, and several evaded reliable monitoring for months after the event.

Several other Briz-M stages have since exploded, producing additional fragments and contributing to a sustained pattern of debris generation from the Briz-M family. The underlying cause has been identified as inadequate passivation of the stage's pressurized propellant systems. The Briz-M was originally designed in the 1990s under operational practices that did not require complete passivation, and residual propellant in spent stages has caused multiple subsequent explosions. Russian launch vehicle operators have reportedly modified the stage design to improve passivation. Still, the population of unpassivated Briz-M stages launched in earlier decades continues to pose a sustained risk of explosion.

By mid-2026, at least five Briz-M stages had experienced post-mission explosions, producing a combined total of more than 2,000 cataloged fragments. The most recent confirmed event occurred in 2019. Each explosion occurred years after the stage's original mission, reflecting the gradual degradation of propellant containment rather than any immediate operational failure. The Briz-M's explosion history has made it one of the most closely watched families of rocket stages in the current debris catalog, and the dozens of Briz-M stages still in orbit from missions conducted before passivation was improved remain individually monitored by both U.S. and European tracking systems.

The lesson of the Briz-M events is that the legacy population of unpassivated upper stages from earlier decades remains a measurable source of new debris, even as modern stages are designed for full passivation. The historical inventory of explosion-prone stages was extensive: several hundred Russian and earlier Soviet stages, several dozen American stages from the pre-passivation era, and smaller numbers from other launch vehicle families. The ongoing rate of explosions from this inventory adds a few percent per year to the total debris population. The eventual depletion of the legacy inventory, through either explosion or controlled removal, will be a crucial step in stabilizing the long-term environment. Still, the timeline for that depletion extends decades into the future.

The 2007 Fengyun-1C Test

On January 11, 2007, the People's Republic of China conducted a direct-ascent anti-satellite test against its own Fengyun-1C weather satellite. The intercept occurred at an altitude of approximately 865 kilometers (537 miles). It produced more than 3,500 cataloged fragments, with statistical models suggesting tens of thousands of additional smaller fragments below the tracking threshold. Because of the high altitude, the fragments are expected to persist for decades to centuries; approximately 2,800 fragments remain cataloged more than nineteen years after the test.

The Fengyun-1C event is the largest single contributor to the contemporary low Earth orbit debris population. It accounts for one of the largest shares of cataloged fragments at altitudes between 800 and 900 kilometers (497 and 559 miles), the most heavily affected band. The fragments occupy a sun-synchronous orbital plane at approximately 99 degrees inclination, which intersects the orbital planes of many operational satellites at high relative velocities, producing an elevated collision risk that has persisted since the test and will continue for the foreseeable future.

The test was deliberately designed to demonstrate technical capability. The interceptor was a kinetic kill vehicle delivered by an SC-19 ballistic missile, derived from Chinese strategic missile programs. The kill vehicle struck the target at a relative velocity of approximately 8 kilometers per second (5 miles per second), producing a catastrophic impact in the technical sense: both the interceptor and the target were completely fragmented. The fragment count exceeded that of either the 1985 American test or the Soviet co-orbital tests of earlier decades, partly because of the target's higher mass and partly because of the higher relative velocity.

The diplomatic response to the Fengyun-1C test was sharply negative. The United States, several European states, and Japan publicly condemned the test and urged China to refrain from further similar tests. NASA's then-administrator Michael Griffin characterized the test as creating a hazardous environment for spaceflight that would persist for decades. The European Space Agency issued a formal protest. The United Kingdom, Australia, Canada, and Japan joined the condemnation. China's official response acknowledged that the test had been conducted but dismissed the reaction as exaggerated and politically motivated.

The test was credited in later policy literature with catalyzing international debris-mitigation discussions that led to the United Nations COPUOS endorsement of the IADC guidelines in 2007 and to the eventual development of the LTS Guidelines in 2019. China has not conducted a comparable destructive test since 2007, although it has conducted several rendezvous and proximity operations that some Western analysts have characterized as anti-satellite-related. The pattern of post-2007 Chinese behavior is consistent with a strategic decision to refrain from further destructive testing while maintaining or expanding non-destructive anti-satellite capabilities. Whether the restraint is durable and whether China would refrain from destructive testing in a future strategic crisis are important uncertainties regarding the trajectory of the debris environment.

The technical lessons from Fengyun-1C were unambiguous. A single deliberate fragmentation event at moderate-to-high altitude could permanently degrade an orbital regime. No natural process operating on policy-relevant timescales could limit the cumulative consequences of debris from such events. The strategic incentive to demonstrate anti-satellite capability, if acted upon at high altitudes, came at a cost borne by the entire spacefaring community for generations. Most spacefaring states have since internalized these lessons, although the persistence of anti-satellite testing programs in several countries indicates that they are not universally accepted.

The 2008 USA-193 Intercept (Operation Burnt Frost)

On February 20, 2008, the United States destroyed its malfunctioning USA-193 reconnaissance satellite as part of Operation Burnt Frost. Launched in December 2006, the satellite failed shortly after deployment and never reached its operational orbit. By late 2007, it was clear the satellite would reenter uncontrollably, and analysis suggested that residual hydrazine fuel aboard the spacecraft might survive reentry and pose a hazard on the ground. The Bush administration authorized an intercept, carried out from the cruiser USS Lake Erie using a modified SM-3 missile.

The intercept occurred at an altitude of approximately 153 miles (247 kilometers), well below the operational orbits of most other satellites. The choice of altitude was deliberate, intended to ensure that fragments from the intercept would experience strong atmospheric drag and decay rapidly. Approximately 175 fragments were cataloged, and most reentered within months of the event. The deliberate choice of a low altitude reflected an awareness of the consequences of debris, a consideration the Chinese test thirteen months earlier had not displayed.

The American test was also publicly justified on grounds that were markedly different from the Chinese rationale. Whereas the Fengyun-1C test was understood internationally as a demonstration of anti-satellite capability, the USA-193 intercept was characterized as a humanitarian operation to prevent ground hazards posed by the toxic hydrazine fuel the failed satellite was expected to carry during reentry. The U.S. government provided extensive briefings to international partners and conducted the intercept with a high degree of public transparency, including pre-positioning observers from allied nations and providing detailed post-event analysis. The contrast with the secrecy surrounding the Chinese test was deliberate, intended to reinforce the narrative distinction between the two events.

The justification for hydrazine was not universally accepted. Critics argued that the hydrazine in USA-193 would have dispersed harmlessly during reentry, that the risk of casualties from the surviving tank had been largely overstated, and that the intercept's true motive was to demonstrate an anti-satellite capability comparable to China's the year before. The technical analyses supporting the official rationale relied on assumptions about the hydrazine tank's survival that have not been independently verified. The strategic context, occurring shortly after the Chinese test and involving capabilities clearly developed for purposes broader than the specific USA-193 contingency, provided plausible support for the alternative interpretation. The truth is probably somewhere between the two readings: the hydrazine concern was real but probably overstated, and the operational opportunity to demonstrate anti-satellite capability was real but probably not the primary motivation.

The contrast between the two events became a standard example in the debris policy literature of how strategic objectives can be achieved with widely varying environmental footprints. The same kinetic intercept technology, applied at different altitudes, produced fundamentally different long-term debris consequences. Later anti-satellite tests by other nations have, with one exception, followed the lower-altitude precedent rather than the higher-altitude one. The technical lesson that altitude choice is the single most important factor in determining the long-term debris consequences of any deliberate fragmentation has been widely incorporated into the operational thinking of national programs.

The 2009 Iridium-Cosmos Collision

On February 10, 2009, the operational Iridium 33 communications satellite collided with the defunct Russian Cosmos 2251 communications satellite at an altitude of about 789 kilometers (490 miles) over northern Siberia. The closing velocity was about 11.7 kilometers per second (7.3 miles per second). Iridium 33 had a mass of about 700 kilograms (1,543 pounds); Cosmos 2251 had a mass of about 950 kilograms (2,094 pounds). The collision was the first known accidental hypervelocity collision between two intact satellites. It produced about 1,800 cataloged fragments from the two spacecraft, most of which remain in orbit today.

The technical details of the collision deserve careful examination because the event has shaped the conjunction screening infrastructure more than any other. Iridium 33 was an active satellite in the Iridium constellation, deployed in 1997 and operating in a polar orbit with an inclination of approximately 86 degrees. Cosmos 2251 was a Soviet-era communications satellite launched in 1993 and abandoned in 1995 after the failure of its service modules. The two satellites operated at similar inclinations but at slightly different altitudes; their operational orbits periodically brought them into close approach.

The U.S. Space Surveillance Network had identified the close approach in advance, and a conjunction warning had been distributed through the channels then in use. However, the predicted miss distance was reported as several hundred meters, well above the threshold that would have prompted Iridium to perform an avoidance maneuver. The actual collision occurred because the conjunction analysis lacked sufficient precision in the cross-track direction; the predicted miss distance was off by an amount that, in retrospect, fell within the prediction's uncertainty.

The chain of decisions leading to the collision has been extensively analyzed. In 2009, the Space Surveillance Network tracked tens of thousands of objects, with limited resources for high-precision analysis of any single conjunction. Conjunction warnings were issued only when the predicted miss distance fell below a threshold; the Iridium-Cosmos miss distance fell within the warning band but did not trigger the high-precision follow-up analysis that might have revealed the actual collision risk. The probability-of-collision metric, which would have explicitly accounted for position uncertainty, was not yet standard practice for conjunction screening. Iridium had limited operational procedures for evaluating conjunction warnings; the company's general practice was to maneuver only when the warning indicated a high probability of collision, and the February 9 warning did not meet that threshold.

The Iridium-Cosmos event was the most consequential accident of the satellite era and prompted a series of operational and institutional responses that have fundamentally reshaped conjunction screening infrastructure. The Combined Space Operations Center, established in 2009, expanded the conjunction warning service to a much broader range of operators and tightened the uncertainty thresholds at which warnings are issued. Commercial space situational awareness providers began developing independent tracking and analysis capabilities, in part to provide redundancy to the U.S. military's monopoly on conjunction screening. The probability-of-collision metric, rather than the simple miss-distance metric, became standard. Operator practices evolved toward more aggressive avoidance maneuvers. Whether these changes would have prevented the 2009 collision had they been in place at the time is impossible to know. Still, they have measurably reduced the probability of similar events in the years since.

The fragments from the Iridium-Cosmos collision are concentrated in a band of altitudes between approximately 500 and 1,000 kilometers (311 and 621 miles), with the densest concentration near the original collision altitude of 789 kilometers (490 miles). The fragments inherited a velocity distribution from the two parent satellites, producing the spread in altitudes. The fragments' orbital planes are similar to those of the two parent satellites, with an inclination near 86 degrees, which is among the most heavily used for low Earth orbit operations and results in high relative velocities during encounters with sun-synchronous traffic. The collision risk to other satellites in the affected altitude bands has been measurably elevated since 2009, with several near-misses attributable to Iridium-Cosmos fragments documented in the years following the event.

The financial consequences of the Iridium-Cosmos event were severe for Iridium. The destroyed satellite was replaced through normal constellation operations, but the operational disruption and the increased insurance costs that followed were estimated at several tens of millions of dollars. Iridium did not pursue a Liability Convention claim against Russia for the loss, in part because of the difficulty of establishing fault under the convention's standards and in part because of the diplomatic complications of pursuing such a claim against a state with which the United States had broader commercial and strategic relationships. The decision not to pursue the claim has been criticized in retrospect as missing an opportunity to test the Liability Convention's applicability to debris-generated damage. Still, it reflected the practical realities of state-to-state space liability disputes during a period of generally cooperative international relations.

The 2016 Hitomi Breakup

On March 26, 2016, the Japanese X-ray astronomy satellite Hitomi (also known as ASTRO-H) experienced a software failure that led to uncontrolled rotation, eventually exceeding its structural limits and shedding multiple components. The event occurred shortly after the satellite's launch and operational checkout, during a maneuver to acquire its first scientific target. A faulty parameter in the attitude control software caused the satellite to misinterpret its orientation. The resulting commanded maneuvers spun the satellite rather than stopping its motion, and within hours, the rotation exceeded the satellite's structural design limits.

The breakup produced ten cataloged fragments, most of which decayed within a few years. The event was minor compared with the largest collision and anti-satellite events. Still, it was historically significant as the first known fragmentation caused entirely by an attitude control system failure. The technical investigation by the Japan Aerospace Exploration Agency (JAXA) revealed a complex chain of failures: a single flawed software upload was propagated through normal operational channels without detection; the onboard verification systems failed to flag the inconsistency; and the safe-mode response that should have arrested the anomalous rotation had been disabled by an unrelated procedure earlier in the mission timeline.

The lesson from Hitomi was that the reliability of spacecraft software was a key contributor to the risk of debris generation. This factor had been underappreciated in the historical mitigation framework. The 2000s mitigation guidelines focused on physical reliability, tank passivation, structural integrity, and end-of-life disposal capability. They gave less attention to the software systems that increasingly mediated all aspects of spacecraft operations. The Hitomi event prompted reviews of attitude control software practices across the major space agencies and contributed to the development of more stringent software verification requirements for future missions. The 2021 revision of the IADC mitigation guidelines provides more explicit treatment of software reliability than the 2002 original.

The 2019 INTELSAT 29e Breakup

On April 7, 2019, the INTELSAT 29e high-throughput communications satellite, at geostationary altitude, experienced a propellant system failure that led to fragmentation. The satellite had been launched in January 2016 and operated successfully for three years before the event. The proximate cause was a leak in the propellant system, which triggered an explosive reaction when residual propellant came into contact with structural materials at elevated temperatures.

The breakup produced multiple cataloged fragments at geostationary altitudes, a regime that, as discussed in Chapter 3, has otherwise been spared significant fragmentation events. The fragments are essentially permanent on policy-relevant timescales, although the rate of subsequent collisions in the geostationary belt remains low because of the regime's characteristic low spatial density. The event was notable as the first consequential fragmentation of a modern operational geostationary satellite and as a demonstration that the geostationary regime is not immune to the fragmentation events that have shaped low Earth orbit.

The post-event analysis identified design weaknesses in the satellite's propellant management that were not apparent during pre-launch testing. The failure mode involved interactions between propellant pressurization and structural thermal cycling that had not been fully characterized during the qualification campaign. The lesson that even modern, well-tested geostationary satellites can produce fragmentation events under specific failure conditions prompted reviews of propellant system design across the geostationary satellite industry. Whether these reviews will prevent similar events in the future is unclear; the geostationary fleet includes a substantial population of older satellites with less rigorous design verification than INTELSAT 29e, and a comparable failure in any of those older satellites could produce a similar or larger fragmentation event.

The 2019 Mission Shakti Test

On March 27, 2019, India conducted a direct-ascent anti-satellite test under the program designation Mission Shakti. The intercept destroyed the Microsat-R satellite at an altitude of approximately 282 kilometers (175 miles). The choice of altitude was deliberately low, consistent with the 2008 American test rather than the 2007 Chinese test. Approximately 130 fragments were cataloged, and most decayed within a year of the event.

Mission Shakti served both domestic and international purposes, which the technical event itself only partially captured. India had been developing anti-satellite capabilities for several years, and the test demonstrated its operational maturity. Prime Minister Narendra Modi announced the test in a televised address, characterizing it as a strategic achievement and emphasizing India's emergence as the fourth nation with proven anti-satellite capability, after the United States, Russia, and China. The political framing was directed mostly at domestic audiences in the run-up to a national election. Still, it also positioned India strategically in the international debate over anti-satellite capability.

The choice of a low altitude was widely interpreted as a deliberate effort to minimize debris and to position India favorably in the international debate over anti-satellite testing. In later statements, Indian officials emphasized that the test produced minimal long-term debris and contrasted the event with the Chinese 2007 test on this basis. The contrast was technically valid: 130 fragments at 282 kilometers (175 miles), mostly decaying within a year, were far less consequential than 3,500 fragments at 865 kilometers (537 miles) that persisted for decades. The Indian framing implicitly accepted the principle that altitude choice was the relevant criterion for responsibility in anti-satellite testing. This framing had been promoted by the United States after Operation Burnt Frost in 2008.

The international response was more muted than the response to Fengyun-1C, in part because of the lower altitude and faster decay timescale, and in part because India's strategic position, as a nonaligned state developing capabilities to balance Chinese pressure, was more difficult to criticize than China's had been. The U.S. State Department issued a measured response that acknowledged the test while expressing concern about the precedent of additional anti-satellite testing. The European Space Agency and several other agencies issued similar measured responses. China was significantly more critical, characterizing the Indian test as destabilizing in ways that the post-Fengyun-1C international response had not characterized the Chinese test. The asymmetric response patterns illustrate how the politics of anti-satellite testing depend extensively on the strategic relationships between the testing state and the responding states.

The event nonetheless contributed to the international policy debate over anti-satellite testing and added to the growing body of evidence that anti-satellite capabilities could be acquired with widely varying environmental footprints. India has not conducted a follow-on test, although Indian space officials have stated publicly that the capability remains available if strategic circumstances require its use. Whether India would conduct a future test at a higher altitude, in circumstances where the strategic value of demonstrating capability against higher-altitude targets outweighs the debris consequences, remains an open question about the future trajectory of anti-satellite testing.

The 2021 Cosmos 1408 Test

On November 15, 2021, Russia conducted a direct-ascent anti-satellite test against its own defunct Cosmos 1408 satellite. The intercept occurred at an altitude of approximately 480 kilometers (298 miles) and produced more than 1,500 cataloged fragments, along with numerous smaller fragments below the tracking threshold. The altitude was chosen to ensure that a large number of fragments would persist for years without producing centuries-long persistence.

The interceptor was a Nudol direct-ascent anti-satellite missile, derived from Russian ballistic missile defense programs. Nudol had been tested non-destructively on multiple occasions before the November 2021 event; the destructive test against Cosmos 1408 marked the system's first operational use against an actual satellite. Russian official statements characterized the test as a routine demonstration of capability. However, the timing, during a period of escalating tensions with NATO over Ukraine, suggested a broader strategic context. The test occurred several months before the Russian invasion of Ukraine in February 2022, and the strategic signaling likely reflected the broader pattern of Russian assertiveness during that period.

The most immediate operational consequence of the test was its impact on the International Space Station. Shortly after the intercept, the fragment cloud passed near the station's orbit, prompting the seven-person crew to take shelter in their two return spacecraft, a Soyuz and a SpaceX Crew Dragon, as a precaution against potential impacts. The shelter procedure lasted about two hours and was repeated several times over the following days as the fragment cloud passed nearby on subsequent station orbits. The event was the most direct demonstration in many years of the operational risks posed by orbital debris to crewed spaceflight, and it received extensive international media coverage. The fact that the test directly endangered the station's crew, including

American and European astronauts as well as Russian cosmonauts, made the diplomatic optics particularly difficult for Russia.

The diplomatic response was sharply negative. The United States, the European Union, the United Kingdom, Japan, Canada, Australia, and several other states publicly condemned the test. The condemnation was particularly notable because it came amid deteriorating U.S.-Russian relations over Ukraine, and U.S. officials used unusually direct language. NASA Administrator Bill Nelson characterized the test as "reckless and dangerous." Secretary of State Antony Blinken issued a formal statement condemning the test as "irresponsible." The European Space Agency, the U.K. Space Agency, and the Japanese government issued similar formal protests. The international response was more unified than the response to the 2007 Chinese test, reflecting both the more advanced state of international thinking on debris policy and the deterioration in relations with Russia.

The episode directly contributed to the U.S. April 2022 announcement of a unilateral commitment to refrain from destructive direct-ascent anti-satellite testing, a commitment subsequently endorsed by more than 30 other states through United Nations resolutions. The U.S. commitment was a significant policy development. The United States maintained anti-satellite capability and had used it in 2008. Still, the Vice President's announcement of a testing moratorium represented a strategic decision that the operational value of demonstrating that capability did not justify the environmental and diplomatic costs. The announcement was widely understood as an effort to establish a behavioral norm that would discourage further testing by other states. Whether the norm holds over the coming decade is one of the central questions for the trajectory of the orbital debris environment.

The Cosmos 1408 event illustrates the contemporary dilemma of anti-satellite testing. The strategic logic of demonstrating capability persists. The technical capability to test at low altitudes, with reduced debris consequences, exists. Yet the pattern of testing, even when conducted at altitudes that are not catastrophic in absolute terms, generates widespread international opprobrium and contributes incrementally to the long-term degradation of the orbital environment. Whether the developing international norm against destructive testing will hold, and whether further tests will follow the higher- or lower-altitude precedent, remains among the central uncertainties about the trajectory of the debris environment.

The 2022 Long March 6A Breakup

On November 12, 2022, an upper stage of the Chinese Long March 6A launch vehicle exploded at altitudes of approximately 700 to 800 kilometers (435 to 497 miles). The stage had been launched several days earlier and deployed a Yunhai-3 weather satellite into orbit. The explosion produced approximately 350 cataloged fragments, adding to the growing population in the most heavily affected band of low Earth orbit. The event was notable because the Long March 6A is a relatively new vehicle, indicating that passivation practices are not yet uniformly applied across the Chinese launch vehicle family.

The Long March 6A entered service in March 2022 and was designed as an upgraded medium-lift launch vehicle for sun-synchronous and other low Earth orbit missions. The explosive fragmentation of an early-flight stage of a new vehicle suggests that passivation procedures in the operational design were either inadequate or inconsistently implemented. Chinese space officials provided a limited public explanation of the event, and the technical analysis available within the international debris community is based primarily on tracking data rather than on primary engineering documentation.

Subsequent Long March 6A flights have continued, and several have placed Chinese constellation satellites into orbit. It remains unclear whether the passivation issues identified in the November 2022 event have been addressed in operational practice. The pattern of Chinese launch vehicle development has been rapid expansion with mixed attention to debris-mitigation practices, and the Long March 6A event illustrates the ongoing risk that inadequate passivation in newer vehicles will generate new debris, even as operational practices for older vehicles improve.

Other Notable Events

Prime Minister Narendra Modi announced the test in a televised address, calling it a strategic achievement and emphasizing India's emergence as the fourth nation with proven anti-satellite capability, after the United States, Russia, and China. The 2015 NOAA-16 fragmentation occurred on November 25 of that year and involved a defunct U.S. weather satellite launched in 2000 and decommissioned in 2014. The satellite produced approximately 275 cataloged fragments at altitudes near 528 miles (850 kilometers). The cause was identified as battery failure, with overheating triggering an internal explosion that fragmented the spacecraft. The lesson from the NOAA-16 event is that batteries, like propellant tanks, can fail in ways that lead to fragmentation. Battery passivation, the practice of discharging batteries to safe states before mission end, is an essential component of the broader passivation discipline.

The 2019 fragmentation of the defunct Chinese rocket body RCS-201 produced approximately 100 cataloged fragments at altitudes near 497 miles (800 kilometers). The event was attributed to a propellant tank failure in a stage that had been in orbit for years. The pattern of older Chinese stages continuing to explode at intervals has been documented across multiple events and represents a sustained but small contribution to the debris population.

On March 18, 2021, the Chinese ocean reconnaissance satellite Yunhai-1 02 fragmented, producing approximately 50 cataloged fragments. The event has been attributed to a collision with a small fragment, possibly from earlier debris. The specific identity of the colliding object has not been publicly confirmed. Still, the event is particularly relevant as a possible early example of a collision driven by debris fragments rather than by intact satellites, a category of events that the cascade analysis identifies as becoming more important as the debris population grows.

The 2024 Resurs-P1 breakup on June 26 involved a defunct Russian Earth observation satellite that produced approximately 100 cataloged fragments at altitudes near 292 miles (470 kilometers). The cause was attributed to a propellant tank failure. The event was notable because the fragments passed near the International Space Station shortly after the breakup, prompting the crew to take shelter as a precaution. The episode was a reminder that routine fragmentation events involving older Russian satellites continue to pose direct operational risks to crewed spacecraft, and that the legacy population of unpassivated Soviet-era and early

Russian satellites remains a persistent contributor to the debris environment.

The Briz-M and Other Upper Stage Explosions

Upper-stage explosions are a category of fragmentation events that warrant separate attention. Spent rocket upper stages retain residual propellant in pressurized tanks, which, over time, can fail in ways that produce explosions. The mechanism is well understood: thermal cycling causes pressure fluctuations; propellant decomposition produces gases that increase tank pressure; tank materials degrade under combined thermal and chemical stress; and eventually a seal or weld fails catastrophically. The result is an explosion that fragments the stage into hundreds or thousands of pieces.

The historical record includes dozens of upper-stage explosions, predominantly involving Russian and earlier Soviet stages, with fewer involving American, European, and Chinese stages. The Delta upper-stage explosion problem of the 1970s was the first large-scale instance. It prompted the development of passivation procedures, including venting residual propellant after spacecraft separation, which later became standard practice. The Delta-K upper stage, used throughout the 1990s, was the principal subject of these events; multiple Delta-K stage explosions contributed significantly to the historical debris environment before passivation was implemented in the late 1990s.

Modern American and European stages are reliably passivated, and the rate of new explosions from these sources has fallen substantially over the past two decades. The pattern is more mixed for Russian stages: passivation has been adopted in newer designs but has not been retroactively applied to the population of stages launched in earlier decades, which continue to explode at intervals of years to decades. The Chinese launch vehicle family has historically had inconsistent passivation practices, though recent designs have improved. The 2022 Long March 6A event illustrates the persistence of these issues.

The cumulative debris contribution from upper-stage explosions, summed across the historical record, is comparable to the combined contribution of anti-satellite tests and accidental collisions. In this sense, the category is a major component of the overall debris environment, even though no single explosion event matches the absolute size of Fengyun-1C or Iridium-Cosmos. The slow, ongoing rate of new explosions from older stages remains a measurable contributor to the population, and the eventual depletion of the older population, through either explosion or controlled removal, will be an important step in stabilizing the long-term debris environment.

Lessons from the Case Studies

Several themes emerge from the case studies considered together. The first is the dominant role of altitude in determining long-term consequences. Events at low altitudes produce debris that decays within years, whereas events at higher altitudes produce debris that persists for decades or centuries. The strategic and operational decisions that determine altitude choice are therefore the single most crucial factor in shaping the environmental footprint of any deliberate fragmentation. The contrast between the 2007 Fengyun-1C event at 865 kilometers (537 miles) and the 2008 USA-193 event at 247 kilometers (153 miles) illustrates this dependence with particular clarity: similar kinetic intercept events at different altitudes produced fundamentally different long-term debris environments.

The second theme is the role of technical maturity in mitigating accidental events. The 2009 Iridium-Cosmos collision occurred because the conjunction-screening infrastructure at the time was inadequate to provide reliable warnings; an equivalent event today would be very unlikely because the screening infrastructure has matured considerably. The Briz-M explosions occurred because passivation was not standard practice; modern stages are unlikely to produce equivalent events. The Hitomi breakup occurred because attitude control software was not adequately verified; modern verification standards address the specific failure mode that produced Hitomi. Technical maturity, in other words, has greatly reduced the rate of accidental fragmentation over the past two decades. However, the rate of legacy events, fragmentations of older systems designed under earlier standards, will continue for years.

The third theme is the persistent role of strategic considerations in driving deliberate fragmentation. The pattern of anti-satellite testing, as reflected in Fengyun-1C, Mission Shakti, and Cosmos 1408, reflects strategic decisions by national programs that the international policy framework has not adequately constrained. The development of the international norm against destructive testing, beginning with the 2022 American commitment, is the most far-reaching policy development on this front in recent years. Still, the norm is fragile, and its persistence cannot be taken for granted. Whether China and Russia, the two leading spacefaring states that have not endorsed the norm, refrain from further destructive testing in the coming years will strongly affect the trajectory of the debris environment.

The fourth theme is the cumulative nature of the debris environment. No single event has produced a catastrophic outcome by itself. The Iridium-Cosmos collision did not trigger a cascade. The Fengyun-1C test did not render its altitude band unusable. But the cumulative effect of the major events, together with the ongoing flow of smaller events and the operational releases of mission-related debris, has brought the most heavily affected altitude bands to or near the cascade threshold. The future trajectory of the environment depends, in this sense, less on any single decision than on the accumulated pattern of all decisions. The policy frameworks discussed in later chapters must be designed to address that cumulative character.

The fifth theme, less often discussed but increasingly visible in the recent record, is the international and political dimension of fragmentation events. The diplomatic response to a fragmentation event depends largely on the strategic relationship between the testing or affected state and the broader international community. The asymmetric responses to the 2007 Chinese test, the 2019 Indian test, and the 2021 Russian test reflect underlying patterns of international relations more than the technical characteristics of the events themselves. In this sense, the political dynamics of debris policy are a continuation of broader international politics by other means, and the success of debris policy depends on integrating debris considerations into the broader fabric of international relations.

Figure 5.1. Altitude Profile of Major Fragmentation Events

(Marker size = number of fragments produced)

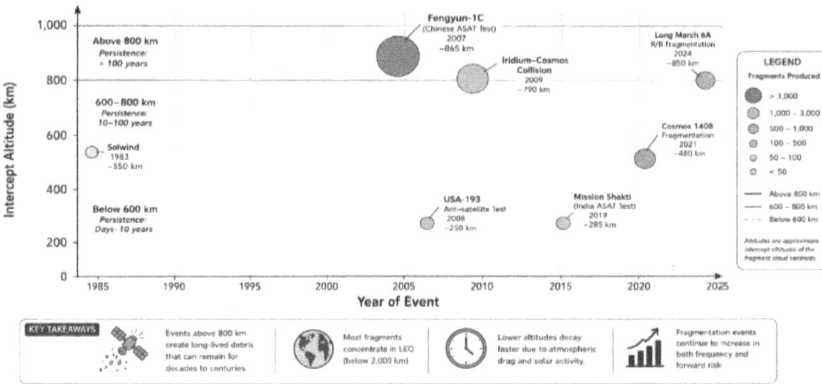

Figure 5.2. Time Evolution of the Iridium 33 / Cosmos 2251 Fragment Cloud, 2009–2026

(789 km altitude; clearance via atmospheric drag)

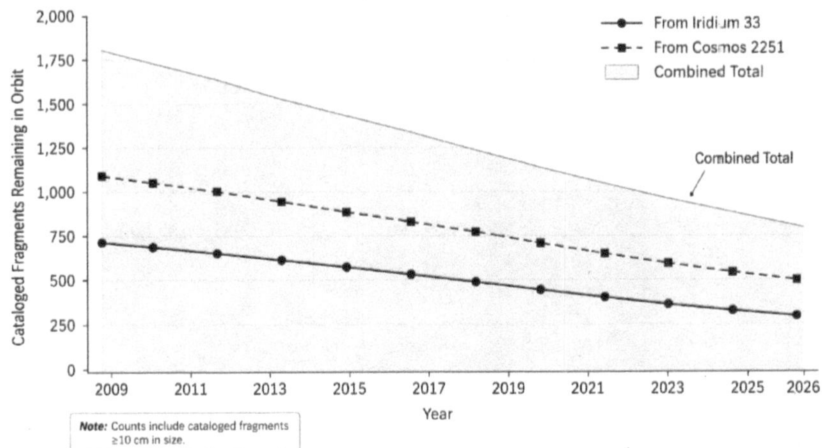

Year	Event	Altitude	Fragments Cataloged	Type
1965	Transit 4A booster	~700Km ~435Miles	~480	Accidental explosion
1985	Solwind P78-1 (ASM-135)	555Km 345 Miles	285	U.S. ASAT test
2007	Fengyun-1C (SC-19)	865Km 537Miles	3,500+	Chinese ASAT test
2008	USA-193 (Burnt Frost, SM-3)	247Km 154Miles	~175	U.S. intercept
2009	Iridium 33 / Cosmos 2251	789Km 490Miles	~1,800	Accidental collision
2015	NOAA-16	~850Km ~528Miles	~275	Battery explosion (post-EOL)
2016	Hitomi (ASTRO-H)	~575Km ~357Miles	~10	Attitude control failure
2019	Microsat-R (Mission Shakti)	282Km 175Miles	~130	Indian ASAT test
2019	INTELSAT 29e	GEO 35,786Km 22,236Miles	~30	Propulsion/battery anomaly
2021	Cosmos 1408 (Nudol)	480Km 298Miles	1,500+	Russian ASAT test
2022	Long March 6A upper stage	700–800 km 435-497 miles	~350	Accidental breakup

Table 5.1. Major Fragmentation Events of the Satellite Era, 1957–2025

Date	Country	System	Target	Altitude	Fragments
1985	USA	ASM-135	Solwind P78-1	555Km 345Miles	285
2007	China	SC-19	Fengyun-1C	865Km 537Miles	3,500+
2008	USA	SM-3 (Burnt Frost)	USA-193	247Km 153Miles	~175
2019	India	Mission Shakti	Microsat-R	282Km 175Miles	~130
2021	Russia	Nudol	Cosmos 1408	480Km 298Miles	1,500+

Table 5.2. Anti-Satellite Test History

Chapter Summary

The orbital debris environment has been shaped primarily by a small number of discrete fragmentation events, the most consequential of which are the 2007 Fengyun-1C anti-satellite test, the 2009 Iridium-Cosmos collision, and the 2021 Cosmos 1408 anti-satellite test. Together with a long history of upper-stage explosions and a smaller number of additional collision and breakup events, including the 1985 Solwind test, the Soviet co-orbital tests of 1968-1982, the 2008 USA-193 intercept, the 2019 Mission Shakti test, the 2016 Hitomi breakup, the 2019 INTELSAT 29e fragmentation, the 2022 Long March 6A breakup, and various NOAA, Briz-M, and Resurs events, these case studies define the historical pattern of debris generation. The lessons they collectively teach, about altitude choice, technical maturity, strategic incentives, international politics, and cumulative effects, shape the policy and operational responses examined in the chapters that follow. The case studies also show that the orbital debris environment is not the inevitable byproduct of routine spaceflight but the product of specific decisions made by specific actors at specific times, and that better decisions in the future could measurably improve the trajectory of the environment.

Chapter 6. Reentry Hazards and Atmospheric Effects

Introduction

The orbital debris problem is conventionally framed as a problem of objects in orbit. But every object in orbit eventually reenters, either through the slow decay caused by atmospheric drag or through deliberate disposal maneuvers, and the consequences of reentry have become an important concern in their own right. The conventional treatment of reentry has focused on the hazard posed by surviving fragments to people and property on Earth's surface, a concern that has been addressed for decades through spacecraft design standards, controlled reentry practices, and the casualty risk thresholds embedded in national licensing frameworks. A newer concern, gaining increasing attention only in the past several years, is the chemical effect of reentered spacecraft material on the upper atmosphere, including the metals and oxides released as spacecraft components vaporize during reentry, and the implications of those releases for atmospheric chemistry, ozone, and climate.

This chapter examines both aspects of the reentry problem. It begins with the physics of reentry, the mechanisms by which spacecraft disintegrate during descent through the atmosphere, and the factors that determine which components survive to ground impact. It examines the casualty risk assessment frameworks used by national space agencies to evaluate planned and unplanned reentries. It surveys the largest uncontrolled reentries of the satellite era, identifying the events that have shaped public understanding and policy responses. It examines the recent emergence of atmospheric impact as a serious debris-related concern, drawing on National Oceanic and Atmospheric Administration (NOAA) measurements of spacecraft-derived metals in stratospheric aerosols, as well as the broader implications for atmospheric chemistry. It examines the policies and design practices that mitigate both categories of reentry impact, including controlled reentry practices and design changes that reduce surviving mass and harmful atmospheric emissions.

The Physics of Reentry

A spacecraft reenters Earth's atmosphere at orbital velocity, typically between seven and eleven kilometers per second (four-point-three and six-point-eight miles per second), depending on the orbit from which reentry begins. The spacecraft's kinetic energy must be dissipated during descent, primarily through aerodynamic heating as it compresses the air ahead of it, converting kinetic energy into thermal energy. The intensity of the heating depends on the spacecraft's velocity, the atmospheric density at each altitude, the spacecraft's geometry, and the materials from which it is constructed.

For typical spacecraft, reentry becomes measurable at about 120 kilometers (75 miles). The atmospheric density at this altitude is about one ten-billionth of sea-level density. Even at orbital velocity, this thin atmosphere produces enough heating to start breaking the spacecraft apart. As the spacecraft descends, atmospheric density increases roughly exponentially, with each 10-kilometer (6.2-mile) decrease in altitude roughly doubling the density. The heating rate increases correspondingly, and at altitudes between 80 and 100 kilometers (50 and 62 miles), the heating becomes sufficient to vaporize most low-melting-point materials.

Components made of low-melting-point materials, such as aluminum, magnesium alloys, plastics, and organic composites, begin to disintegrate at altitudes between 80 and 100 kilometers (50 and 62 miles). Aluminum, the principal structural material of most spacecraft, has a melting point of approximately 660 degrees Celsius (1,220 Fahrenheit) and begins to vaporize at temperatures above approximately 2,470 degrees Celsius (4,479 Fahrenheit). Aerodynamic heating and the convective transport of molten material away from the spacecraft surface combine to produce rapid mass loss for aluminum structures during the reentry breakup phase. Magnesium alloys, with even lower melting points, disintegrate similarly. Plastics and organic composites typically pyrolyze and burn at lower temperatures and are completely consumed during reentry.

Components made from higher-melting-point materials, such as stainless steel, titanium alloys, certain ceramic composites, and refractory metals, survive longer and may reach lower altitudes before disintegrating. Titanium, with a melting point of approximately 1,670 degrees Celsius (3,038 degrees Fahrenheit), can withstand lower temperatures than aluminum. Stainless steel, with melting points typically between 1,400 and 1,500 degrees Celsius (2,552 and 2,732 degrees Fahrenheit), withstands similarly high temperatures. Tungsten and other refractory metals, used in specific applications such as radiation shielding, can withstand ground impacts even during high-velocity reentries. Carbon composite materials used in heat shields are engineered to ablate gradually and may survive atmospheric reentry in significant quantities.

A small fraction of the original spacecraft mass typically survives to ground impact. The surviving mass depends on the spacecraft's design, trajectory, and the materials from which it is constructed. For typical low Earth orbit satellites, the surviving fraction is between 10 and 40 percent of the original mass, distributed across multiple fragments ranging from grams to tens of kilograms. The largest surviving components are typically dense, compact items: pressure tanks, reaction wheels, batteries with steel cases, and similar hardware. Lighter, more diffuse components, such as solar arrays, antennas, and communications electronics, are typically completely consumed.

NASA's Debris Assessment Software, used to assess reentry casualty risk for U.S. missions, predicts the surviving fragments from each spacecraft component using empirical models calibrated to observed reentry events. The predictions are inherently uncertain; reentry dynamics are chaotic at fine detail, and small variations in initial conditions produce substantially different outcomes. However, the statistical distribution of surviving fragments is sufficiently well characterized to support meaningful risk assessment. The software requires detailed input data: the mass, materials, and geometry of each spacecraft component, as well as the initial reentry conditions. The output is a distribution of expected surviving fragments and the associated casualty area for each fragment.

The Mechanics of Atmospheric Heating

The aerodynamic heating that drives reentry breakup arises from the compression of air ahead of the spacecraft as it travels at hypersonic speeds. The compressed air is heated to extremely high temperatures, typically several thousand degrees Celsius (Fahrenheit) for low Earth orbit reentries, and the heat is transferred to the spacecraft surface through both convective and radiative mechanisms. The heating rate depends on the local atmospheric density, the spacecraft's velocity, and its geometry; it is most intense at the spacecraft's leading edge and at points of high curvature, where the bow shock is strongest.

The temperature of the compressed air can be estimated from the spacecraft's velocity using the energy balance between kinetic and thermal energy in the shock layer. At a velocity of 7.7 kilometers per second (4.8 miles per second), characteristic of low Earth orbit, the stagnation temperature in the shock layer is approximately 26,000 Kelvin, far above the melting point of any spacecraft material. The actual temperature at the spacecraft surface is lower than the stagnation temperature due to heat-transfer dynamics. Still, it remains high enough to vaporize most materials with sufficient exposure time.

The duration of intense heating is typically short, on the order of one to two minutes, after which either the spacecraft has descended into denser atmospheric layers, where it has been completely decelerated, or it has disintegrated to the point that the surviving fragments individually have low ballistic coefficients and decelerate rapidly. The total energy deposited in the spacecraft during this brief interval is enormous: the kinetic energy of a 1,000-kilogram (2,204.6 pounds) spacecraft at orbital velocity is approximately 30 gigajoules, equivalent to the energy released by the combustion of approximately 700 kilograms (1,543 pounds) of gasoline.

The radiative component of heating becomes particularly important at very high velocities, such as those encountered during returns from cislunar or interplanetary trajectories. For typical low Earth orbit reentries, convective heating dominates. For Apollo-era lunar return capsules, which entered the atmosphere at approximately 11 kilometers per second (6.8 miles per second) after their translunar return, radiative heating contributed measurably to the thermal load and required heat shield designs different from those used for low Earth orbit return vehicles. The Crew Dragon and Starliner capsules used for ISS crew rotation, returning from low Earth orbit at 7.7 kilometers per second (4.8 miles per second), use heat shield designs optimized for a lower-velocity return profile.

Casualty Risk Assessment

The casualty risk associated with a spacecraft reentry is conventionally expressed as the expected number of casualties per reentry event. NASA's standard, established in NSS 1740.14 and applied throughout the U.S. government, requires that the expected casualty risk for any planned reentry not exceed 1 in 10,000. The threshold is a policy choice, not a derived quantity, and it balances the cost of compliance, whether by building spacecraft with surviving mass below the threshold or by designing them for controlled reentry over open ocean, against the public risk posed by unplanned or uncontrolled reentries. It surveys the largest uncontrolled reentries of the satellite era, identifying the events that have shaped public understanding and policy responses.

The 1-in-10,000 threshold has significantly influenced spacecraft design over the past two decades. Spacecraft designed for low Earth orbit that are not equipped for controlled reentry must demonstrate, through analysis, that their surviving fragments will not pose a casualty risk above the threshold. The principal approaches to compliance include reducing surviving mass through design choices, using lower-melting-point materials, eliminating dense components that survive intact, and breaking up large structures into smaller pieces that disintegrate more thoroughly, or implementing a controlled reentry capability that allows the operator to direct the reentry to an area with sufficiently low population density so that any surviving fragments will fall in unpopulated regions.

For spacecraft that comply by design, the analysis must show that the cumulative casualty area of all surviving fragments, weighted by the probability of impact across Earth's populated regions, yields an expected casualty count below the threshold. The casualty area of a fragment is the area within which a person standing in the open would be injured or killed by the impact, calculated from the fragment's mass, shape, and impact velocity. NASA's standard practice defines the casualty area as the sum of the fragment's cross-sectional area and that of a standing human, scaled by the impact energy. Population density data, typically drawn from gridded global population databases, weights the casualty area by the probability of impact at each location along the spacecraft's possible reentry footprint.

For spacecraft that comply via controlled reentry, the operator must demonstrate that the controlled reentry capability is sufficiently reliable to ensure that the residual probability of uncontrolled reentry, combined with the casualty risk in the uncontrolled case, remains below the threshold. Controlled reentry is generally the more straightforward path to compliance for large spacecraft. However, it requires the spacecraft to retain attitude control and propulsion capability through the end of its mission, with operational consequences that not all mission designs can accommodate. The propellant required for a controlled reentry burn from low Earth orbit typically is 5 to 10 percent of the spacecraft's launch mass, a considerable allocation that affects the feasibility of small or low-cost spacecraft.

Casualty risk treatment varies internationally. ESA has adopted a more stringent internal requirement of 1-in-100,000 for some recent missions, reflecting both the agency's particular sensitivity to debris issues and the political consequences of a casualty event for European space programs. The Japanese space agency JAXA consistently applies the IADC-aligned 1-in-10,000 threshold. The Chinese space program does not appear to apply a comparable internal threshold for its launch vehicle stages, as illustrated by the pattern of Long March 5B reentries discussed below. The Russian space program has historically applied less stringent reentry standards, although recent practices for the Russian Federal Space Agency missions have approached international norms.

ESA formalized its more stringent standard through the Space Debris Mitigation Policy for Agency Projects, most recently revised in 2023, which specifies that ESA-led missions shall comply with a maximum casualty risk of 1-in-100,000 for controlled and uncontrolled reentry events. The higher standard reflects both the agency's particular institutional sensitivity to debris issues and the practical reality that ESA missions tend to operate at altitudes and with orbital parameters that allow the more stringent threshold to be met through design rather than requiring an expensive controlled reentry capability.

The casualty risk threshold itself has been the subject of recurring debate. A 1-in-10,000 threshold per reentry event implies that, given the 2024 reentry rate (approximately 2,000 per year), the expected number of reentry casualties would be about one every five years. The actual observed rate is far lower than this expected value; no confirmed reentry casualties have been recorded, although several near misses have occurred. The discrepancy reflects both that not all reentries are subject to the same stringency standards and that the threshold itself is a conservative upper bound rather than a precise prediction. Some analysts have argued that the threshold should be tightened as the volume of reentries grows; others have argued that the threshold represents a reasonable balance and should be maintained.

Notable Uncontrolled Reentries: Skylab and Kosmos 954

The historical record includes several uncontrolled reentries that have drawn widespread public attention and prompted policy responses. The 1979 Skylab reentry was the first widely publicized example. NASA had hoped to use the upcoming Space Shuttle program to reboost and preserve Skylab's orbit. However, delays in Shuttle development pushed the effort beyond the point at which reboost was feasible, and Skylab reentered the atmosphere uncontrolled in July 1979. The station, with a mass of approximately 85 tons (77 metric tonnes), was the largest single object to undergo uncontrolled reentry up to that point.

NASA and ground observers worldwide closely tracked Skylab's reentry. The predicted reentry point shifted as solar activity altered atmospheric drag, and the station's orientation changed unpredictably during its final orbits. NASA attempted to use small attitude maneuvers to influence the reentry point, but control authority was limited, and predictions remained uncertain until the final hours. Debris from the reentry fell across the Indian Ocean and western Australia, in a band stretching from south of Perth eastward across the Australian outback. The largest surviving fragments weighed several hundred kilograms each, including a section of the airlock and an oxygen tank that survived nearly intact. No injuries or significant property damage resulted, and the most famous artifact of the event was a $400 fine (Australian dollars) levied by the Western Australian town of Esperance against NASA for "littering," a fine that was paid in 2009 by a California radio personality.

The Skylab episode caused public alarm at the time and contributed to subsequent policies requiring more careful planning for the disposal of large spacecraft. The Skylab reentry was the proximate cause of the IADC's later requirement that large spacecraft be designed for controlled reentry, and it shaped disposal planning for subsequent space stations. The eventual reentry of the Russian Mir space station in March 2001 was conducted under controlled disposal, with engine burns that directed the station's descent into the South Pacific Ocean. The same approach is planned for the eventual disposal of the International Space Station around 2030, with NASA contracting SpaceX to develop a U.S. Deorbit Vehicle to perform the controlled reentry burn.

The 1978 Kosmos 954 reentry, several months before Skylab, drew less media attention but had more serious consequences. Kosmos 954 was a Soviet ocean-reconnaissance satellite powered by a nuclear reactor for radar imaging of the sea. The satellite suffered a control-system failure and reentered uncontrolled over northern Canada on January 24, 1978, depositing radioactive debris across a swath of the Northwest Territories. The cleanup operation, conducted jointly by Canadian and American teams under Operation Morning Light, took approximately 10 months and recovered numerous radioactive fragments scattered across an area of approximately 124,000 square kilometers (77,051 square miles).

The cleanup costs were estimated at approximately $14 million in 1978 dollars (equivalent to roughly $65 million in 2026 dollars). The Soviet Union paid Canada $3 million in compensation under the 1972 Liability Convention in a 1981 settlement. The settlement did not constitute a formal admission of liability but provided partial financial compensation for cleanup costs. It was the only formal application of the Liability Convention to a debris-related event and established a precedent that has been cited in subsequent discussions of similar events.

Kosmos 954 remains the most consequential single reentry event of the satellite era, primarily because of the radioactive contamination it caused. The Soviet Union and later Russia continued to use nuclear reactors on a small number of military satellites for several years after the Kosmos 954 incident. Still, the practice was sharply curtailed in the 1980s and effectively ended after the Cold War. The legacy of nuclear-powered satellites includes several others that were eventually uncontrolled.

However, in those cases, the radioactive material was either successfully contained by design features (such as boost-to-storage-orbit maneuvers that placed the reactors in long-lived orbits) or dispersed at altitudes high enough that natural decay timescales extended beyond the time required for radioactivity to decay to safe levels.

Tiangong-1 and the Long March 5B Series

The 2018 reentry of the Chinese Tiangong-1 space station, decommissioned in 2016 and abandoned without a controlled reentry capability, sparked months of public concern as the predicted reentry footprint shifted over populated regions worldwide. Tiangong-1 had been launched in 2011 as China's first space station and had hosted multiple crewed missions before being placed in standby mode in 2013. Communications with the station were lost in March 2016, and the station was effectively abandoned, leaving it unable to perform controlled disposal.

The actual reentry on April 2, 2018, occurred over the South Pacific Ocean. No surviving fragments were recovered, and no damage or casualties resulted. The episode nonetheless contributed to the international policy debate over uncontrolled reentries of large spacecraft. Western space surveillance networks extensively tracked the Tiangong-1 reentry, and the predicted reentry corridor was published openly, allowing the public to follow the descent. The transparency of the tracking, however, did not change the fact that the reentry was uncontrolled and that the impact location was determined by atmospheric drag and orbital geometry rather than by deliberate operator action.

The most consequential pattern of uncontrolled reentries in recent years has involved the Chinese Long March 5B launch vehicle. The Long March 5B is a heavy-lift rocket whose core stage is approximately 30 meters (98 feet) long and weighs approximately 23 tonnes (25 tons) empty, making it one of the largest objects ever to enter low Earth orbit. The vehicle has been used to launch components of the Tiangong space station, beginning with the Tianhe core module in 2021. The rocket's operational design leaves the core stage in low Earth orbit, where it reenters uncontrolled.

The first Long March 5B mission, in May 2020, launched a prototype Chinese crew capsule. The core stage reentered uncontrolled on May 11, 2020, leaving fragments along a track that crossed Africa from north to south. Fragments were recovered in Côte d'Ivoire, where they damaged buildings in several villages but caused no casualties. The Côte d'Ivoire fragment recovery was the first publicly documented case of significant property damage from an uncontrolled reentry in many years. It drew intense international criticism of the Chinese launch program.

The second mission, in April 2021, launched the Tianhe core module of the Tiangong space station; the core stage reentered the atmosphere uncontrolled on May 9, 2021, with fragments falling into the Indian Ocean. The third mission, in July 2022, launched the Wentian module. The core stage reentered uncontrolled on July 30, 2022, with fragments falling in Southeast Asia and the Indian Ocean; several fragments were recovered in Borneo. The fourth mission, in October 2022, launched the Mengtian module; the core stage reentered the atmosphere uncontrolled on November 4, 2022, with fragments falling into the Pacific.

The Long March 5B reentries drew sustained international criticism. After the May 2021 reentry, NASA Administrator Bill Nelson issued a strongly worded statement, characterizing the event as a failure to meet "responsible standards." Subsequent reentries prompted similar statements from NASA, ESA, and several other agencies. The U.S. Space Command provided detailed tracking information for each reentry, contrasting U.S. transparency with the limited information from the Chinese government. The Chinese government's response was generally dismissive, characterizing the criticism as anti-Chinese sentiment rather than a legitimate technical concern.

Subsequent versions of the Long March 5B family have reportedly been modified to enable controlled reentry. However, no public confirmation has been provided, and the operational practices for future Tiangong-related missions remain unclear. The Long March 5B variant used in 2025 for additional Chinese missions has not produced uncontrolled reentries comparable to those of 2020-2022, suggesting that operational practices have changed. However, these changes have not been publicly documented.

The pattern of Long March 5B reentries highlights a recurring tension in international debris policy. The IADC mitigation guidelines and the UN COPUOS-endorsed framework recommend controlled reentry for large spacecraft, but compliance is voluntary, and there is no enforcement mechanism. National practices vary widely. Several leading spacefaring states, including the United States and the European Space Agency, have adopted internal requirements that effectively prohibit the launch of designs that produce uncontrolled reentries above the casualty risk threshold. China, Russia, and several other states have not adopted equivalent internal requirements. This variation largely reflects the absence of a binding international standard and the reluctance of leading powers to subject their launch programs to such external constraints.

Recent Uncontrolled Reentries and Near-Miss Events

Beyond the Long March 5B series, the past several years have seen several uncontrolled reentries of smaller objects that have nonetheless attracted attention. The May 2024 reentry of a SpaceX Crew Dragon trunk section, with fragments recovered in North Carolina, sparked heightened public discussion about the risk of casualties posed by commercial space activities. The trunk section, which is jettisoned before crew capsule reentry and remains in orbit until atmospheric decay clears it, was not part of the protected casualty risk framework that applies to the crew capsule itself. The surviving fragments illustrated that the design choice to deorbit the trunk separately produced a casualty risk that the controlled disposal of the crew capsule did not.

The March 2024 reentry of a discarded International Space Station battery pallet produced fragments recovered in Florida, including one that penetrated the roof of a private home. The pallet had been jettisoned from the ISS in 2021 as part of a battery replacement campaign, with the expectation that it would burn up entirely on reentry, but the actual reentry produced surviving fragments that predictive modeling had not anticipated. The recovered Florida fragment was approximately 700 grams (1.5 pounds) of metal that had entered through the roof of the home and embedded itself in the floor. No injuries resulted, but the event was the most direct demonstration in many years of the casualty risk posed by uncontrolled reentries, and it prompted reviews of the disposal practices for ISS-discarded equipment.

The pattern of these recent events, small individual reentries from a wide range of operators, each with limited individual casualty risk but cumulatively contributing to the overall reentry-related risk, illustrates the limits of an event-by-event approach to casualty risk assessment. The 1-in-10,000 per-event threshold was established when the rate of uncontrolled reentries was modest and cumulative casualty exposure was small. As the rate has increased with the deployment of large constellations and their associated regular satellite disposals, cumulative exposure has correspondingly increased. Whether the per-event threshold remains adequate, or whether a cumulative-exposure threshold should also apply, is one of the open questions in the current reentry policy.

Atmospheric Effects of Reentries

A topic that has attracted renewed attention in recent years is the chemical effects of reentering spacecraft on the upper atmosphere. The conventional view of reentry has focused on the fate of large surviving fragments and the casualty risk they pose. However, the materials that vaporize during reentry do not simply disappear; they enter the upper atmosphere as gases and small particles, where they can persist for years and influence atmospheric chemistry and physics in ways that have only recently begun to be quantified.

The most important recent finding comes from a 2023 paper by Daniel Murphy and colleagues at the National Oceanic and Atmospheric Administration (NOAA), published in the Proceedings of the National Academy of Sciences (Murphy et al., 2023). The paper reported direct measurements of stratospheric aerosols collected by aircraft at altitudes around 18 kilometers (11 miles) in the polar stratosphere. The aerosols contained measurable concentrations of aluminum, lithium, copper, and other metals that are not naturally present at those altitudes in those proportions. The isotopic signatures and elemental ratios were consistent with the expected products of spacecraft reentry rather than meteoritic infall. The paper estimated that approximately 10 percent of the measured aerosol particles contained metals derived from reentered spacecraft.

The methodology behind the NOAA measurements is worth noting because it provides the empirical foundation for subsequent analyses. The measurements were taken using NASA's WB-57 high-altitude aircraft, which carried a particle analysis mass spectrometer capable of identifying the elemental composition of individual aerosol particles at the picogram scale. The instrument, called the PALMS (Particle Analysis by Laser Mass Spectrometry), can analyze tens of thousands of individual particles per flight and produce a statistical picture of the aerosol composition at the sampled altitudes. The 10 percent figure for spacecraft-derived metals was based on an extensive sample size and reflects a robust statistical signal.

The implications of this finding are still being assessed. Stratospheric aerosols play an important role in both ozone chemistry and climate. Sulfate aerosols, naturally present and augmented by volcanic emissions, are known to affect ozone depletion and climate forcing. Aerosols of metallic origin from spacecraft reentries differ in composition. They may have distinct effects, but the scale of their contribution, which is growing rapidly with the increasing rate of satellite reentries, is now large enough that the question cannot be ignored. Several follow-on studies are underway as of mid-2026, and the field is developing rapidly.

Aluminum oxide is the most important specific compound produced during spacecraft reentry from a chemical standpoint. Aluminum is the principal structural material of most spacecraft, and its oxidation during reentry produces aluminum oxide particles that enter the stratosphere directly. Aluminum oxide is known to participate in heterogeneous chemistry on stratospheric particles, including reactions that can affect ozone concentrations. Historically, aluminum oxide in the stratosphere has come predominantly from emissions from solid rocket motors during launches, a source that has been studied since the 1990s. The new contribution from satellite reentries is currently smaller than the launch-phase contribution. Still, it is growing more rapidly and is projected to overtake it within the next decade if current trends continue.

The rate of atmospheric metal injections from reentries is expected to rise rapidly over the coming decade. The deployment of large satellite constellations, with disposal timescales of five to seven years, implies that the steady-state mass of satellite material reentering each year will eventually reach hundreds of tonnes (tons). At Starlink's projected operational scale of approximately 30,000 satellites with five-year average lifetimes, approximately 6,000 satellites would reenter each year, with a combined mass of approximately 1,500 to 2,000 tonnes (1,653 to 2,205 tons). Other large constellations would add to this total. The cumulative annual mass entering the upper atmosphere from constellation reentries alone could approach 5,000 tonnes (5,512 tons) by the early 2030s, an order of magnitude greater than the historical baseline.

To put the scale in concrete terms: in 2020, the total annual mass of satellites reentering the atmosphere was about 200 metric tons. By 2025, that figure had grown to roughly 400-600 metric tons, driven primarily by the early turnover of Starlink Generation 1 satellites. At Starlink's projected full operational scale of 30,000 or more satellites with five-year average operational lifetimes, the steady-state annual reentry mass from that single constellation alone approaches 1,500 to 2,000 metric tons. Adding the projected contributions from Amazon Leo, OneWeb, and the Chinese Guowang and Qianfan constellations brings the total projected annual reentry mass above 4,000 metric tons by the early 2030s. This is more than twenty times the 2020 baseline. The 2023 NOAA measurements were approximately 200 metric tons annually. Whether the atmosphere's chemistry responds linearly to a twentyfold increase in metallic input, or whether threshold effects produce disproportionate consequences at higher loading rates, is among the most consequential open questions in contemporary atmospheric science.

The full implications of this material input for atmospheric properties remain unknown. Ozone depletion is a specific concern: aluminum oxide and other metallic compounds can catalyze ozone-destroying reactions, as documented for sulfate aerosols, and this may extend to new compounds from spacecraft reentries. Climate effects are another concern: stratospheric aerosols can alter Earth's energy balance by interacting with both incoming solar radiation and outgoing thermal radiation. The nature of the climate effect depends on aerosol size, composition, and altitude distribution, and metallic aerosols from spacecraft may have optical properties different from those of natural sulfate aerosols. Whether the net effect is warming or cooling, and whether it is large enough to be appreciable relative to other anthropogenic climate forcings, is currently under active research.

Solid Rocket Motor Emissions

A related topic is the atmospheric impact of solid rocket motor firings. Solid rocket motors, used as boosters on many launch vehicles, emit large quantities of aluminum oxide particles during their burn. These particles are deposited in the stratosphere along the launch vehicle's ascent trajectory and can persist for years. The cumulative atmospheric effect of solid rocket motor emissions has been studied since the 1990s. It is generally considered small compared with other anthropogenic atmospheric perturbations, but it is nonzero and has been included in environmental analyses for major launch programs.

Solid rocket motors contain relatively high concentrations of aluminum. Aluminum powder is commonly added to solid propellants as a fuel and typically accounts for about 16–18 percent of the total propellant mass. A typical solid rocket booster, such as those used on the Space Shuttle or the Atlas V launch vehicle, contains hundreds of tonnes (tons) of propellant and emits tens of tonnes of aluminum oxide during its burn. The historical Space Shuttle program produced approximately 80 tonnes (88 tons) of aluminum oxide per launch from the two solid rocket boosters; cumulative emissions from 135 Shuttle launches contributed substantially to the historical anthropogenic aluminum oxide loading of the stratosphere.

Modern launch vehicles differ in how they use solid propellant. The Falcon 9, used for most Starlink launches, uses liquid propellants throughout and produces no aluminum oxide emissions. Atlas V, Delta IV, and other vehicles with solid rocket boosters produce significant emissions per launch. Ariane 5 and Ariane 6 use solid rocket boosters, contributing to Europe's share of stratospheric aluminum oxide. The trend toward liquid-only launch vehicles, driven by economic and reusability considerations, has incidentally reduced per-launch aluminum oxide emissions. However, the rising total launch rate has partially offset this reduction.

The growing rate of launch traffic, particularly the high-cadence launches associated with mega-constellations, has prompted renewed attention to launch-phase atmospheric impacts. Reusable launch vehicles, which avoid first-stage disposal, may also have different atmospheric footprints than expendable vehicles. The full life-cycle atmospheric impact of a typical launch, including manufacturing, propellant production, launch-phase emissions, and reentry of expended stages, is now being studied as a coherent environmental question rather than a series of separate issues. The conclusions are tentative but generally point to a need for more careful environmental analysis of launch programs as their volume grows.

Black Carbon and Other Combustion Products

Beyond aluminum oxide, launch vehicle exhaust introduces other materials into the stratosphere with potential atmospheric implications. Black carbon, a fine soot produced by the incomplete combustion of hydrocarbon fuels, is emitted by all liquid-fueled launch vehicles using kerosene or methane propellants. The black carbon emissions from a single launch are small compared to the natural atmospheric burden. Still, black carbon in the stratosphere has different properties from that in the lower atmosphere: it absorbs solar radiation effectively, contributing to direct atmospheric heating, and its long residence time in the stratosphere amplifies its per-mass climate impact relative to surface emissions.

A 2022 study by researchers at the National Center for Atmospheric Research projected that launch traffic from planned mega-constellations could produce stratospheric black carbon emissions sufficient to contribute to climate forcing within the next decade measurably. The projected emissions were small compared to total anthropogenic climate forcing, on the order of one percent or less, but they were not negligible. The study identified the need for improved characterization of launch-vehicle exhaust composition and for developing life-cycle environmental analyses that integrate launch emissions with other space-related environmental impacts.

The interaction between launch emissions and stratospheric chemistry is more complex than a simple additive model would suggest. Emissions are deposited at specific altitudes during ascent, and chemical species vary along the trajectory. The deposited materials interact with existing stratospheric chemistry over months to years before removal. Some interactions are direct: nitrogen oxides emitted at high altitudes can catalyze ozone depletion. Others are indirect: heating of stratospheric air by absorbing aerosols can alter circulation patterns, thereby affecting ozone transport. The full chemistry of these interactions is the subject of active research, with both modeling and measurement programs working to characterize the system.

Light Pollution and Reentry Visibility

A minor but visible aspect of the reentry problem is the visibility of bright reentries to ground observers. The reentry of a massive spacecraft produces a bright streak across the sky, often visible for tens of seconds and spanning hundreds of kilometers. Several recent reentries, including those of Falcon 9 second stages, retired satellites, and the Long March 5B core stages, have been observed by large numbers of people and have generated widespread social media attention. The visibility itself is harmless, but these events have brought the reentry phenomenon into public awareness in a way that the technical and policy literature has not.

Falcon 9 second-stage reentries are particularly distinctive because of the rocket's design. After delivering its payload to orbit, the second stage performs a deorbit burn that returns it to the atmosphere along a controlled trajectory. The reentry produces a bright fireball visible across broad regions of the U.S. and the Pacific. SpaceX has published reentry trajectories in advance for several missions, allowing observers to anticipate and photograph the events. Falcon 9 reentries are among the most reliable examples of controlled reentry practice in modern operations and serve as a public demonstration of the controlled-disposal alternative to the uncontrolled approach used by some other launch programs.

For ground-based astronomy, reentries pose a distinct concern. A bright reentry passing through a telescope's field of view can saturate detectors and disrupt observations. These events are infrequent and, individually, have modest impact, but their cumulative impact, given the increasing rate of reentries, is non-zero. The astronomical community has begun to track reentry events and incorporate them into the broader analysis of human impacts on ground-based observation. Chapter 15 examines the broader question of the astronomical impacts of space activity in detail.

Controlled Reentry as Mitigation

The primary mitigation for both casualty risk and atmospheric impact is controlled reentry. A controlled reentry directs the spacecraft to a designated impact zone, typically the South Pacific Ocean Uninhabited

Area (SPOUA), also known as Point Nemo. This region, centered at approximately 48 degrees south latitude and 124 degrees west longitude, is the most remote location on Earth from any inhabited landmass; the nearest inhabited territory, Pitcairn Island, is more than 2,500 kilometers (1,553 miles) away. Over the past several decades, more than 270 spacecraft have been deliberately reentered into the SPOUA, including Mir, the first Tiangong, multiple Progress and HTV cargo vehicles, and many smaller spacecraft.

The SPOUA approach effectively addresses casualty risk but not atmospheric impact, because the materials vaporized during reentry enter the upper atmosphere regardless of where the surviving fragments ultimately impact. Atmospheric effects can be reduced only by reducing the mass reentered, the altitude at which most of the breakup occurs, or the chemical composition of the reentering material. None of these is straightforward to address through controlled reentry alone, and the long-term solution will likely require either an active debris removal capability that captures spacecraft before reentry or completely different spacecraft designs that produce less harmful reentry products.

The IADC and ISO standards on controlled reentry require that the casualty risk for any planned reentry stay below the 1-in-10,000 threshold and that the controlled reentry capability remain available through the end of the mission. ESA has adopted a more stringent internal requirement of 1-in-100,000 for some recent missions. The U.S. Federal Aviation Administration's Office of Commercial Space Transportation imposes equivalent requirements on licensed commercial launches and reentries. The combined effect of these national-level requirements has measurably improved the safety of planned reentries over the past two decades. However, variation in national practice means that some launching states continue to produce reentries that would not meet the most stringent standards.

Controlled reentry requires that the spacecraft retain functional propulsion, attitude control, and communications capability throughout the deorbit maneuver. The deorbit burn is typically performed at a specific point in the orbit to direct the descent toward the intended impact zone, with the precise timing and duration of the burn determining the impact location. For low Earth orbit spacecraft, the deorbit burn typically consumes propellant equal to approximately 5 to 10 percent of the spacecraft's launch mass and lasts several minutes. The propellant must be reserved for this purpose throughout the entire mission, with operational margins to ensure availability.

Design for Demise

A complementary approach to reducing casualty risk is "design for demise," the deliberate selection of spacecraft materials and configurations to maximize the fraction of mass that disintegrates during reentry. The approach has been adopted by some satellite manufacturers, particularly for satellites operating at altitudes where natural decay clears them within the disposal timeline. By replacing dense, high-melting-point components with lower-melting-point alternatives where feasible, using aluminum housing instead of stainless steel where structural requirements permit, and using plastic structures instead of metal where electromagnetic shielding is not required, the surviving mass can be substantially reduced.

The design-for-demise approach is most applicable to small- to medium-sized satellites in low Earth orbit. For larger spacecraft, structural requirements typically constrain material choices, limiting the achievable reduction in surviving mass. For spacecraft operating at high altitudes where natural decay does not occur within the disposal timeline, controlled reentry remains the only practical compliance path, regardless of design-for-demise effort. The combination of design-for-demise and controlled reentry capability provides the most robust compliance with casualty risk requirements, and modern spacecraft design has trended toward incorporating both approaches.

A specific challenge in design for demise is managing pressure tanks, batteries, and other dense components that have historically been the principal contributors to the surviving mass. Pressure tanks designed to contain high-pressure gas or liquid throughout the operational mission must withstand structural loads that limit material choices. Batteries use materials with specific electrochemical properties that are not easily replaced. The engineering work required to redesign these components for improved reentry performance is extensive, and the cost has historically been justified only when the alternative, controlled reentry, was not practical for operational reasons.

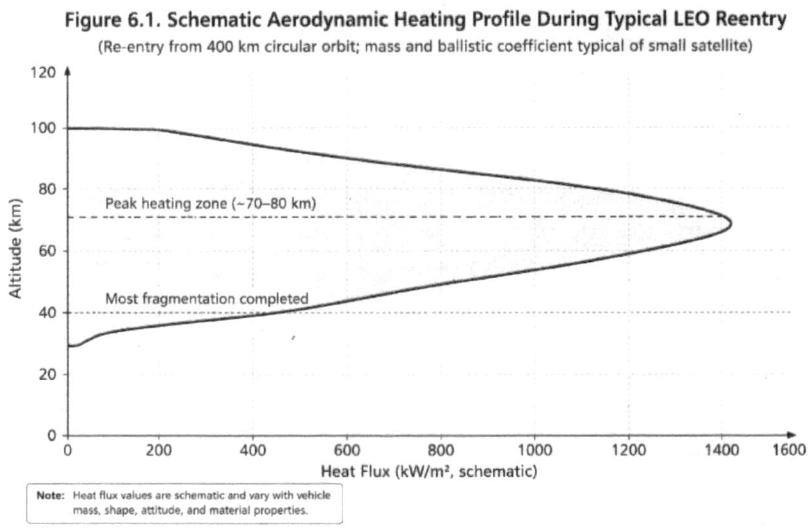
Figure 6.1. Schematic Aerodynamic Heating Profile During Typical LEO Reentry
(Re-entry from 400 km circular orbit; mass and ballistic coefficient typical of small satellite)
Altitude (km)
Heat Flux (kW/m², schematic)
Peak heating zone (~70–80 km)
Most fragmentation completed
Note: Heat flux values are schematic and vary with vehicle mass, shape, attitude, and material properties.

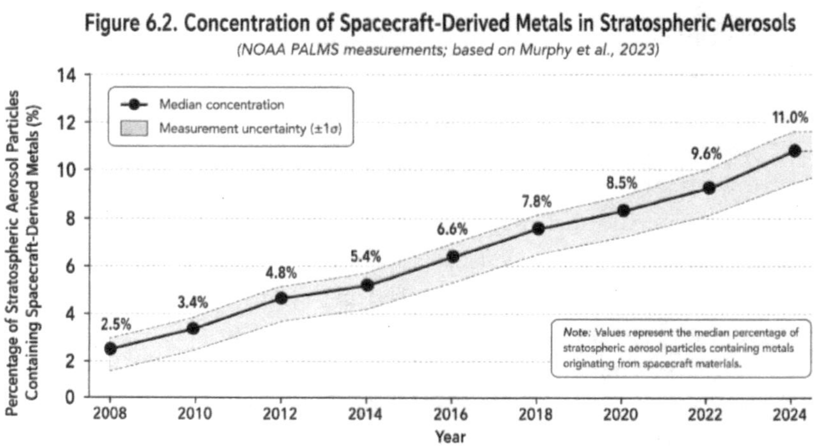
Figure 6.2. Concentration of Spacecraft-Derived Metals in Stratospheric Aerosols
(NOAA PALMS measurements; based on Murphy et al., 2023)
Percentage of Stratospheric Aerosol Particles Containing Spacecraft-Derived Metals (%)
Year
Median concentration
Measurement uncertainty (±1σ)
2.5%
3.4%
4.8%
5.4%
6.6%
7.8%
8.5%
9.6%
11.0%
Note: Values represent the median percentage of stratospheric aerosol particles containing metals originating from spacecraft materials.

Jurisdiction	Threshold	Authority
United States	1 in 10,000	NASA-STD-8719.14, FCC, FAA
European Space Agency	1 in 10,000	ESA Space Debris Mitigation Policy
Japan (JAXA)	1 in 10,000	JAXA Space Activities Act compliance
United Kingdom	1 in 10,000	UK Space Industry Act 2018
France (CNES)	1 in 10,000	Loi sur les Opérations Spatiales 2008
IADC Guidelines	No specific value	Reference to national standards

Table 6.1. Casualty Risk Thresholds Applied by Major Spacefaring States

Year	Object	Mass (tonnes)	Notes
1978	Kosmos 954	~3.8	Soviet RORSAT; radioactive debris in NW Canada
1979	Skylab	~77	Debris in Western Australia
1991	Salyut 7	~40	Argentina
2018	Tiangong-1	~8.5	South Pacific Ocean
2020	Long March 5B core stage	~21	Côte d'Ivoire (May 11)
2021	Long March 5B core stage	~21	Indian Ocean (May 9)
2022	Long March 5B core stage	~21	Borneo (July 30)
2022	Long March 5B core stage	~21	Pacific (November 4)

Year	Object	Mass (tonnes)	Notes
2024	ISS battery pallet (EP-9)	~2.6	Fragment recovered in Florida (March)
2024	Crew Dragon trunk	~1	Fragments in North Carolina (May)

Table 6.2. Notable Uncontrolled Reentries, 1978–2025

Chapter Summary

The reentry of spacecraft and rocket stages from Earth orbit produces two distinct impacts: the risk of casualties from surviving fragments reaching the ground, and the chemical impact of vaporized spacecraft material in the upper atmosphere. The first category has been addressed for decades through spacecraft design standards, controlled reentry practices, and casualty risk thresholds embedded in national licensing frameworks. Compliance has been markedly better in the United States and Europe than in some other leading spacefaring states, with the Long March 5B series providing the most prominent recent example of uncontrolled reentries from a large national launch program. The second category, the atmospheric effects of reentry, has emerged as a growing concern only in the past several years, driven by NOAA measurements showing significant accumulations of spacecraft-derived metals in stratospheric aerosols. The rate of atmospheric metal injections is expected to grow exponentially as the rate of satellite reentries from large constellations increases, and the implications for ozone, climate, and other atmospheric properties are now under active study.

Both categories of impact reinforce the broader argument that orbital debris is not solely a problem of objects in orbit but one with consequences extending to Earth's surface and atmosphere. Mitigation strategies for the two categories overlap but are not identical: controlled reentry effectively addresses casualty risk but not atmospheric impact, whereas design-for-demise approaches reduce both surviving mass and the materials that vaporize. The integrated treatment of reentry as a debris-related environmental issue, rather than an isolated operational concern, is a growing emphasis in contemporary debris policy. The chapters that follow examine the legal, regulatory, technical, and operational frameworks within which these mitigation strategies operate, as well as the broader questions of governance and responsibility that the cumulative environmental impact of space activity raises.

Chapter 7. International Space Law and Policy

Introduction

The international legal framework governing space activity rests on a small number of multilateral instruments developed during the Cold War and a larger body of subsequent guidelines, principles, and norms that have emerged through the United Nations Committee on the Peaceful Uses of Outer Space and other forums. By the standards of contemporary international law, the framework is surprisingly thin. The five core treaties, the Outer Space Treaty, the Rescue Agreement, the Liability Convention, the Registration Convention, and the Moon Agreement, were drafted before most of the contemporary debris problem existed and addressed it only obliquely. The substantive rules governing debris generation and management have been developed primarily through non-binding guidelines, with the IADC mitigation guidelines and the UN COPUOS Long-term Sustainability Guidelines as the central documents.

This chapter examines the international framework in detail. It begins with the five core treaties, identifying the legal principles each establishes and the limitations of each as applied to debris. It examines the IADC, COPUOS, and ITU as the principal forums for ongoing international coordination. It examines the LTS Guidelines as the most authoritative recent statement of international consensus. It examines the Artemis Accords as a significant recent addition to the framework. It identifies the principal gaps in the current framework and the prospects for addressing them. The treatment is sufficiently detailed to support the analysis in subsequent chapters of national regulatory frameworks (Chapter 8) and of the economics and insurance of space activity (Chapter 13), both of which build on the international foundation established here.

The Outer Space Treaty

The Treaty on Principles Governing the Activities of States in the Exploration and Use of Outer Space, including the Moon and Other Celestial Bodies, was opened for signature on January 27, 1967, and entered into force on October 10, 1967. It is the foundational instrument of space law. As of mid-2026, the treaty has been ratified by 115 states, including all major spacefaring states, and has been signed but not yet ratified by an additional 22. The treaty's principal substantive provisions establish that outer space is not subject to national appropriation by claim of sovereignty; that space activities must be conducted for the benefit of all humanity; that nuclear weapons and other weapons of mass destruction may not be placed in orbit or stationed in space; that astronauts are envoys of mankind and must be assisted by all states; that states are responsible for their national activities in space, whether conducted by governmental or non-governmental entities; and that states are liable for damage caused by their space objects.

The treaty does not directly address debris. The drafters in 1967 were concerned with broader questions, including the militarization of space, the prevention of celestial appropriation, and the establishment of a framework for the peaceful exploration that the early space age seemed to promise. The cumulative environmental consequences of space activity, which would emerge as central to space governance fifty years later, were not anticipated. The treaty's provisions, therefore, apply to debris only by extension: the requirement that states bear responsibility for the activities of their national operators provides the legal basis for national debris regulations, and the requirement that space activities be conducted for the benefit of humanity is sometimes invoked to argue that debris-generating activities are inconsistent with the treaty's broader principles.

The treaty's principles on state responsibility are especially important for debris policy. Article VI of the treaty establishes that states are internationally responsible for "national activities in outer space," whether conducted by governmental agencies or by non-governmental entities, and that the activities of non-governmental entities require authorization and continuing supervision by the appropriate state. This principle has shaped the licensing frameworks that nearly all leading spacefaring states have adopted for commercial space activities. Without the treaty's establishment of state responsibility, the development of national licensing frameworks for commercial debris mitigation would have lacked the international legal foundation that lends them legitimacy.

Article VII complements Article VI by establishing the principle of state liability for damage caused by space objects. Fleshed out in detail by the 1972 Liability Convention, the principle holds that the launching state is financially responsible for damage caused by its space objects. Together, Articles VI and VII create a regulatory chain: states are responsible for the activities of their national operators, including damage caused by those operators' space objects, and therefore have an incentive to regulate national activities to prevent damage that would expose them to liability. This regulatory chain is the basic mechanism by which the international legal framework translates into national debris regulations.

The treaty's most important practical consequences for debris are indirect. The principle of state responsibility holds that any commercial debris-generating activity, such as a satellite explosion or a constellation that fails to dispose of its members, is, in legal terms, the activity of the launching state, which is responsible for the consequences. The principle of state liability creates incentives for launching states to regulate the debris-generating practices of their national operators, though the strength of those incentives varies widely across jurisdictions. The principle of free access to space, established in Article I of the treaty, provides the legal basis for international cooperation in space surveillance and tracking, which is foundational to debris management.

A key tension in the treaty's relationship to debris policy concerns whether debris itself qualifies as a "space object" under the treaty's definitions. Article VIII provides that a state retains "jurisdiction and control" over space objects it has launched or that are launched from its territory. If debris fragments are space objects under this definition, the launching state retains jurisdiction over them indefinitely, creating legal complications for active debris removal by entities other than the launching state. The IADC and other coordination bodies have generally treated debris fragments as space objects under the treaty, meaning that any active removal mission targeting fragments produced by another state's space activities requires either the launching state's consent or another legal basis. These legal complications have been recognized but not fully resolved and remain an active issue in current debris policy discussions.

The Rescue Agreement

The Agreement on the Rescue of Astronauts, the Return of Astronauts, and the Return of Objects Launched into Outer Space, opened for signature in 1968 and entered into force the same year, addresses two related obligations: the rescue and assistance of astronauts in distress and the return of space objects to their launching states. The agreement was developed during the early space age, when both the United States and the Soviet Union were conducting human spaceflight programs, and the possibility of an astronaut emergency on the territory of a third state was a significant concern.

The agreement's provisions on the return of space objects are particularly relevant to debris policy. Article 5 of the agreement requires that any state on whose territory a space object is found, or that recovers a space object from areas beyond national jurisdiction, return the object to the launching state upon that state's request. The provision applies to all space objects, including debris that survives reentry and reaches Earth's surface. In principle, fragments from uncontrolled reentries, such as the Long March 5B fragments recovered in Côte d'Ivoire and Borneo, are subject to return under the agreement.

In practice, the return provisions have rarely been invoked. Most surviving fragments from uncontrolled reentries are either too small to warrant recovery, lost in the ocean or in unpopulated areas where they are never found, or retained by the recovering state for its own purposes. The Cosmos 954 fragments recovered in Canada in 1978 were returned to the Soviet Union under the agreement's provisions, providing one of the few formal applications of the return requirements. The agreement does not address compensation for recovery costs, which are addressed separately under the Liability Convention discussed below.

The agreement's relevance to contemporary debris policy is modest. The framework was designed for occasional, isolated reentry events of national significance, not for the growing rate of small-fragment reentries from commercial constellations. Whether the agreement's provisions apply meaningfully to the routine reentry of constellation satellites, or whether an adapted framework should govern the recovery of fragments from such reentries, remains a developing issue in space law. The question is unlikely to be resolved soon, since practical occasions for invoking the agreement remain rare even as the rate of underlying reentries increases.

The Liability Convention

The Convention on International Liability for Damage Caused by Space Objects, opened for signature in 1972 and entered into force the same year, builds on the liability framework established by the Outer Space Treaty. The convention establishes two distinct liability regimes. For damage caused by a space object on the surface of the Earth or to aircraft in flight, the launching state is absolutely liable, that is, liable regardless of fault. For damage caused by space objects in space, the launching state is liable only if fault can be demonstrated.

The absolute liability regime for surface damage has been invoked only once: in the aftermath of the 1978 reentry of Kosmos 954, when Canada pursued a claim against the Soviet Union for the costs of cleaning up radioactive debris in the Northwest Territories. The Soviet Union eventually paid $3 million, less than the Canadian claim of $14 million, but the case set a precedent that has influenced subsequent space-debris policy. Several other reentry events, including various Long March 5B reentries, have sparked public discussion of potential Liability Convention claims, but no claims have been formally pursued. The 2024 Crew Dragon trunk fragment recovered in North Carolina and the 2024 ISS battery pallet fragment that struck a Florida home both occurred on U.S. territory and involved U.S. launching states, eliminating the international dimension that would have triggered the Liability Convention.

The fault-based liability regime in space has repeatedly posed challenges. In the 2009 Iridium-Cosmos collision, both Iridium and Cosmos 2251 were launched by states (the United States and Russia, respectively). However, the question of fault was complicated by the fact that Cosmos 2251 was a defunct satellite that had been an uncontrolled object for years. Iridium ultimately did not pursue a liability claim, in part because demonstrating fault was difficult and in part because the diplomatic complications of pursuing a claim against Russia outweighed the financial recovery that any claim might have produced. The case effectively established that the fault-based liability regime is inadequate to address debris-generated damage in many practical scenarios.

The convention's definition of "damage" is broad, encompassing loss of life, personal injury, or other impairment of health, as well as loss of or damage to property belonging to states, persons (natural or juridical), or international intergovernmental organizations. The breadth of the definition is intended to ensure that all reasonable categories of damage can be addressed under the convention. In practice, the limitations have been less in the definition of damage than in the procedural requirements for pursuing claims, the difficulty of attributing damage to specific space objects, and the political costs of state-to-state liability disputes.

The convention's procedures require that claims be presented through diplomatic channels, with negotiation between the claimant state and the launching state as the first step. If negotiation does not produce a settlement within one year, the claim can be referred to a Claims Commission established under the convention's procedures. The Claims Commission consists of three members, one chosen by each of the two states involved and a chairman chosen jointly. The procedural requirements have been criticized as cumbersome, particularly for modest-scale damage, where the diplomatic costs of pursuing the claim may outweigh the financial recovery. No Claims Commission has ever been convened, and the only formal application of the convention, the 1981 Kosmos 954 settlement, was reached through bilateral negotiation rather than through the formal Commission procedures.

The Liability Convention's limitations as a debris-management instrument are significant. The convention applies only to damage that has occurred, not to debris generation that has not yet caused damage. It provides no mechanism for the international community to constrain debris-generating practices that have not yet produced specific harm. It places the burden of proof on the claimant, which is challenging in cases of space-based damage, where attribution and fault are technically difficult to establish. And it operates on a state-to-state basis, which limits its utility for the growing share of space activity conducted by commercial operators across multiple jurisdictions. Reform of the convention has been discussed in academic and policy literature for decades, but has not produced substantive changes; the convention remains essentially the same as it was in 1972.

The Registration Convention

The Convention on Registration of Objects Launched into Outer Space was concluded in 1975 and entered into force on September 15, 1976. The convention requires launching states to maintain national registries of space objects and to provide information about their launches to the United Nations, which maintains a public register accessible through the UN Office for Outer Space Affairs. The required information includes the launching state's name, an appropriate designator for the space object, the date and territory or location of the launch, basic orbital parameters, and the space object's general function.

The convention provides the foundation for the international tracking of space objects. It is the basis for the U.S. Space Surveillance Network and equivalent national tracking systems in organizing their catalogs. However, the information provided under the convention is generally minimal: orbital parameters are typically launch-epoch rather than current, functions are described only in general terms, and updates over the satellite's operational lifetime are not required. As a result, the UN register is not a comprehensive operational database that would support detailed debris-management activities; for that purpose, operators rely on national tracking services and commercial data providers.

The convention's primary practical limitation is the lack of registration requirements for debris itself. A spacecraft must be registered at launch, but the catalog of fragments produced by a fragmentation event is not registered separately. The relationship between catalog entries and the original launching state is sometimes ambiguous, particularly for older debris. The IADC and other coordination bodies have developed informal practices for tracking debris and attributing it to specific events. Still, the formal registration framework remains tied to the original launches rather than to the resulting populations of debris.

A specific issue with the convention is the inconsistent application of its requirements. Many states report launch information faithfully and promptly; others report with lengthy delays or with limited information. China, for instance, has historically reported launch information with relatively limited operational detail compared to the United States and ESA. Russia's reporting has been variable, with some launches reported promptly and others not registered for extended periods. As a result, the UN register, while important as a record of state declarations of launch activity, is neither a complete nor a current operational catalog of space objects.

The convention has not been meaningfully updated since it entered into force in 1976, and its limitations have become more apparent as the volume and complexity of space activity have grown. The deployment of large satellite constellations, with thousands of individual satellites operated by a single entity, has strained the registration framework's focus on individual objects. The increasing role of commercial operators has complicated the relationship between the registered launching state and the operational entity controlling the satellite. The development of in-orbit servicing and active debris removal capabilities, in which one space object operates on or removes another, raises questions about how these operational changes should be reflected in registration. None of these issues has been formally addressed through amendment of the convention. However, the LTS Guidelines and other recent instruments have begun to address some of them through non-binding guidance.

The Moon Agreement

The Agreement Governing the Activities of States on the Moon and Other Celestial Bodies, opened for signature in 1979 and entered into force in 1984, is the least ratified of the five core space treaties. As of mid-2026, only 18 states have ratified the agreement, and none of the major spacefaring states, including the United States, Russia, China, Japan, France, or India, has done so. The agreement establishes that the Moon and other celestial bodies constitute the "common heritage of mankind" and that an international regime should be established to govern the exploitation of their natural resources.

The common heritage principle is the primary reason the leading powers have not ratified the agreement: it implies a shared regime for resource benefits that conflicts with the commercial space resource activities several leading powers, particularly the United States, have begun to pursue. The U.S. Commercial Space Launch Competitiveness Act of 2015 explicitly authorizes U.S. citizens to engage in the commercial recovery and use of space resources, a position the U.S. government considers consistent with the Outer Space Treaty's prohibition on national appropriation but inconsistent with the Moon Agreement's common heritage framework. Luxembourg, the United Arab Emirates, and other states have adopted similar national legislation. Japan has adopted equivalent legislation. None of these states is a party to the Moon Agreement.

The Moon Agreement's relevance to orbital debris is limited. The agreement focuses on celestial bodies, including the Moon, asteroids, planets, and similar objects, rather than on the orbital regime around Earth. Its environmental protection provisions, including the principle that states shall avoid harmful contamination and adverse changes to the environment of celestial bodies, could, in principle, be extended to orbital debris in lunar or cislunar space. Still, the agreement's limited ratification makes such an extension primarily theoretical.

The Moon Agreement's principal practical relevance to current debris policy is that it serves as a cautionary example of the limitations of multilateral treaty-making in space governance. The agreement was developed through extensive multilateral negotiation but ultimately failed to secure ratification from the states whose participation would have been most consequential. The pattern of major spacefaring states pursuing alternative frameworks, such as the Artemis Accords, national resource legislation, and bilateral arrangements, rather than amending or extending the Moon Agreement, illustrates the difficulty of achieving a comprehensive international agreement on contested space governance issues.

The Inter-Agency Space Debris Coordination Committee

The Inter-Agency Space Debris Coordination Committee (IADC) was established in 1993 as a forum for national space agencies to coordinate debris research and develop common technical standards. The founding members were NASA, ESA, the Russian Space Agency, and the Japanese space agency. Subsequent expansion has brought membership to thirteen agencies, including the China National Space Administration, the Indian Space Research Organization, the Korean Aerospace Research Institute, the Italian Space Agency, the German Aerospace Center, the

French space agency CNES, the United Kingdom Space Agency, the Canadian Space Agency, and the Ukrainian space agency.

The IADC operates through working groups that address specific technical topics: measurements, the environment and databases, protection, mitigation, and active debris removal. Each working group comprises technical experts from member agencies and produces consensus documents reflecting the participating agencies' technical agreement. The IADC has no legal authority and creates no binding obligations; its outputs are technical recommendations that member agencies can adopt voluntarily and that often serve as the basis for subsequent national or international policy.

The committee's most significant output is the Space Debris Mitigation Guidelines, originally published in 2002 and revised in 2021. The guidelines establish the technical baseline for spacecraft design and operations to limit debris generation, including requirements for end-of-life disposal, passivation of upper stages, prevention of intentional breakups and collisions, and limitation of mission-related debris. The 2002 version established the twenty-five-year disposal rule, which became the international baseline for the next two decades; the 2021 revision tightened several requirements, addressed issues that had emerged since the original publication, and incorporated specific guidance for large constellations.

The IADC guidelines have shaped national and international debris policy more than any other single technical document. The United Nations Committee on the Peaceful Uses of Outer Space adopted a set of aligned guidelines in 2007, formalizing the IADC technical work as international policy. The International Organization for Standardization developed ISO 24113, a standard on space systems debris mitigation that closely aligns with the IADC guidelines and provides a framework for compliance assessment. Most leading national space agencies have adopted internal mitigation requirements that closely follow the IADC framework. Commercial operators, particularly those licensed in jurisdictions that have adopted the IADC framework, design their spacecraft to comply with the guidelines.

The IADC's effectiveness as an international coordination mechanism depends heavily on member agencies' willingness to align their national practices with the committee's consensus outputs. Willingness has been high across most agencies, particularly in the United

States, Europe, and Japan. It has been more variable among the Chinese and Russian agencies, both of which participate in the committee but maintain national practices that diverge from the consensus in specific areas. The Long March 5B uncontrolled reentries discussed in Chapter 6 are one example of Chinese practices that diverge from the IADC framework; the continued explosions of legacy Russian upper stages are another. The committee's design as a voluntary technical coordination body, without enforcement authority, limits its capacity to address divergence among member agencies.

United Nations Committee on the Peaceful Uses of Outer Space

The United Nations Committee on the Peaceful Uses of Outer Space (COPUOS) was established by the United Nations General Assembly in 1959 and serves as the principal international forum for developing space governance policy. The committee operates through two subcommittees: the Scientific and Technical Subcommittee, which addresses technical matters, and the Legal Subcommittee, which addresses legal matters. The full committee meets annually to review the subcommittees' work and adopt recommendations for action by the General Assembly.

COPUOS played a central role in developing the five core space treaties during the 1960s and 1970s. Each treaty was negotiated by COPUOS and adopted by the General Assembly on the committee's recommendation. The committee's treaty development has slowed considerably since the Moon Agreement failed to attract widespread ratification, and it has not developed a major new treaty in more than four decades. Instead, it has shifted toward developing non-binding instruments, principles, guidelines, and declarations that can address contemporary issues without the sustained multilateral negotiation that treaty-making demands.

The committee's most consequential recent output is the Guidelines for the Long-term Sustainability of Outer Space Activities, adopted in 2019 after several years of negotiation. The guidelines, examined in detail below, address a wide range of sustainability issues, including debris mitigation, conjunction screening, the safety of nuclear power sources in space, and capacity-building for emerging spacefaring states. The committee has also produced the 2007 Space Debris Mitigation Guidelines, which are derived from the IADC guidelines, as well as a series of other documents addressing specific aspects of space governance.

The committee's effectiveness has been the subject of considerable analysis. On the one hand, it provides the only universal forum for space governance discussions and has been the site of extensive multilateral consensus-building for over six decades. On the other hand, its work has slowed since the 1970s, and its outputs have been predominantly non-binding rather than treaty-level. The committee's structure, with all decisions taken by consensus, has been criticized for enabling individual states to block progress that the broader membership would support. However, the consensus rule also ensures that adopted instruments have broad legitimacy.

Among the committee's emerging areas of work in 2026 is addressing the growing pace of commercial space activity within an institutional framework designed for state-led space programs. The committee's procedural rules give voice to states but not directly to commercial operators, even though commercial operators now conduct much of today's space activity. Various proposals for adaptation have been discussed, including formal observer status for industry associations and consultative arrangements that would allow commercial input into specific technical discussions. None of the proposals has been adopted yet, and the committee continues to operate primarily as a state-to-state forum.

Long-term Sustainability Guidelines

The Guidelines for the Long-term Sustainability of Outer Space Activities, adopted by COPUOS in 2019, are the most comprehensive recent statement of international consensus on space governance. The document contains 21 guidelines organized into four categories: the regulatory framework for space activities; the safety of space operations; international cooperation, capacity-building, and awareness; and scientific and technical research and development. Developed through more than five years of multilateral negotiation, the guidelines represent the most extensive consensus document on space governance since the LTS treaty discussions of the 1970s.

The guidelines cover several debris-relevant areas. Guideline A.4 addresses the supervision of national space activities, requiring states to ensure that their national operators comply with applicable international obligations. Guideline B.1 addresses the provision of conjunction warnings, encouraging states to share orbital data and coordinate conjunction screening services. Guideline B.4 addresses the long-term performance of space objects, including end-of-life disposal and passivation. Guideline B.5 addresses orbital debris monitoring, encouraging the development and sharing of measurements of the debris environment. Guideline B.6 addresses the safety of space operations from a debris perspective, including conjunction avoidance and the prevention of intentional fragmentation.

The guidelines are not binding like a treaty. They are recommendations that states are encouraged to apply voluntarily through their national licensing frameworks and operational practices. Framing the guidelines as voluntary recommendations was a deliberate negotiating choice; binding obligations would have required treaty-level processes, which were considered unlikely to succeed in the political environment of the late 2010s. The guidelines' voluntary character limits their direct effect on debris-generating practices but does not eliminate it: they provide a normative reference point against which national practices can be evaluated, and they have shaped the development of national licensing frameworks across multiple jurisdictions.

The implementation of the LTS Guidelines has been uneven since their adoption. Some leading spacefaring states, including the United States, members of the European Union, Japan, and the United Kingdom, have incorporated the guidelines' principles into their national frameworks. Others have partially implemented the guidelines or have not implemented them in any documented form. The Long March 5B uncontrolled reentries, which occurred after the guidelines' adoption, illustrate the limited effectiveness of voluntary norms when they conflict with specific operational practices that states are unwilling to change.

The committee continues to refine and extend the guidelines. A working group has been established to develop additional guidelines addressing topics the 2019 document did not fully cover, including specific aspects of large-constellation operations, the regulation of commercial debris-removal services, and the governance of cislunar activity. The work remains ongoing as of 2026, with additional guidelines expected to be adopted in the coming years.

International Telecommunication Union

The International Telecommunication Union (ITU) is the United Nations specialized agency for information and communication technologies. Its role in space governance is more focused than that of COPUOS. Still, no less important: the ITU coordinates the use of the radio-frequency spectrum and orbital slots for geostationary satellites. Without ITU coordination, the use of radio frequencies for space communications would be subject to chaotic interference, and the assignment of geostationary slots would be subject to disputes that the international system would be unable to resolve.

The ITU's role in debris policy is important but indirect. The agency does not directly regulate debris generation, but its coordination of orbital slots and frequencies provides leverage that national regulators have begun to use to enforce debris compliance. The U.S. Federal Communications Commission, examined in detail in Chapter 8, has linked its frequency licenses to debris compliance through the September 2022 five-year disposal rule. Other national regulators have followed similar approaches, using ITU coordination obligations as a lever to impose debris-related requirements.

The ITU has begun to address debris-related questions more directly through its World Radio Communication Conferences, which convene every three to four years to update the international Radio Regulations. The 2023 World Radio Communication Conference adopted several decisions on large constellations, including provisions for the equitable sharing of frequencies among non-geostationary constellations and procedural requirements for coordinating new constellation deployments. These provisions do not directly address debris generation, but they shape the regulatory environment in which constellation operators design and deploy their systems.

The geostationary slot allocation framework, under which the ITU assigns specific longitude slots to national administrations, has been the subject of recurring debates over equity. Developing countries have argued that the historical allocation of slots favored early spacefaring states and that the framework should be adapted to provide more equitable access to the geostationary regime. The debate has not produced fundamental changes to the allocation framework but has shaped the procedural rules governing new allocations. The geostationary regime's debris environment, examined in Chapter 3, depends on operators' discipline in vacating slots at the end of life; the ITU's allocation framework, by determining who has rights to which slots, indirectly affects patterns of disposal and graveyard placement.

The Artemis Accords

The Artemis Accords, opened for signature in October 2020, are a significant recent addition to the international space governance framework. The accords are not a treaty in the formal sense; they are a non-binding statement of principles to which signatory states commit voluntarily. As of mid-2026, more than fifty states have signed the accords, including the United States, the United Kingdom, Japan, Australia, Canada, the United Arab Emirates, Italy, France, Brazil, India, the Republic of Korea, and many others. Russia and China have not signed and have stated that they consider the accords incompatible with their preferred multilateral frameworks.

The accords set out ten principles, including peaceful purposes, transparency, interoperability, emergency assistance, registration of space objects, the release of scientific data, the preservation of historic sites, the use of space resources, the deconfliction of activities, and orbital debris. The orbital debris principle commits signatories to follow IADC and COPUOS standards for debris mitigation and to share information on debris-generating events. The principle is not new in substance, but its incorporation into the accords creates a normative framework in which adherence to the IADC and COPUOS standards is part of a broader package of cooperative space governance.

The accords have been characterized in different ways by different observers. Supporters describe them as a flexible, scalable framework that can be expanded incrementally as space activity grows and that complements the existing treaty framework. Critics describe them as an alternative to multilateral treaty processes that fragment international space governance into U.S.-led versus non-U.S.-led blocs. The truth is probably somewhere between the two readings: the accords do offer a useful coordinating framework for the growing community of states engaged in lunar and cislunar activity, and they do represent a different model from the treaty-based approach of the 1960s and 1970s. Whether the model proves durable and whether it eventually produces a consensus that the multilateral treaty processes have not remained to be seen.

The accords' specific provisions on debris reflect the broader evolution of debris policy over the past several decades. The commitment to follow IADC and COPUOS standards is unremarkable, as most leading spacefaring states already make similar commitments through their domestic regulatory frameworks. The provision for sharing information about debris-generating events extends existing joint warnings and data-sharing practices to a broader set of cooperative arrangements. Whether the provisions produce substantive changes in debris-generating practices depends on whether signatories implement the commitments through their domestic frameworks. Early evidence suggests that implementation has been strongest in the United States, Europe, and Japan, and more variable in other signatory states.

Bilateral and Regional Arrangements

Beyond the multilateral framework, a broad set of bilateral and regional arrangements addresses specific aspects of space cooperation, including debris-related issues. The most important bilateral arrangements have historically been between the United States and Russia, including agreements governing the operation of the International Space Station and bilateral data-sharing arrangements that support conjunction screening. The deterioration of U.S.-Russian relations after 2014, particularly after the 2022 Russian invasion of Ukraine, has strained these arrangements but has not eliminated them. Operational cooperation on the ISS has continued despite the deterioration of broader relations, and the principal data-sharing arrangements have remained intact.

Regional arrangements include the European Space Agency's cooperative framework, which coordinates space activities among its 22 member states; the Asia-Pacific Space Cooperation Organization, which coordinates among China and several smaller Asian states; and the

African Space Agency, established in 2018 with headquarters in Cairo. Each of these organizations addresses debris-related issues to varying degrees within its broader cooperative framework. ESA has the most developed debris policy framework among regional organizations, with internal mitigation requirements that are among the most stringent worldwide.

A specific category of bilateral arrangement that has expanded in recent years is space situational awareness data-sharing agreements between the United States and other states. The United States, through Space Command, has agreements with more than 30 other states to share conjunction warnings and tracking data. The agreements vary in scope but generally permit the U.S. to issue conjunction warnings for satellites operated by the partner state in exchange for various forms of cooperation. The expansion of these agreements has profoundly influenced the international SSA infrastructure over the past decade. It has positioned the United States as the central node of a global conjunction-screening network.

Gaps in the Current Framework

The international legal and policy framework for space, despite extensive development over six decades, still has significant gaps that affect debris policy. The most consequential gap is the absence of binding international rules governing debris generation. The IADC and COPUOS guidelines are voluntary, and their effective implementation depends on individual states' willingness to incorporate them into binding national frameworks. States that choose not to do so face no international consequences beyond reputational costs. The Long March 5B reentries are the clearest recent example of behavior that violates the spirit of the international guidelines. Yet they have not produced any concrete consequences for the launching state.

A second gap is the Liability Convention's inadequacy as a debris-management instrument. The convention applies only to damage caused, not to the generation of debris. It places the burden of proof on the claimant and requires cumbersome diplomatic processes for routine debris-related damage. The convention has been formally invoked only once in fifty years, despite the growing number of debris-generating events during the same period. Reform of the convention has been discussed but has not produced substantive change.

A third gap is the lack of a framework for active debris removal by entities other than the launching state. The Outer Space Treaty's principle of state jurisdiction over space objects, combined with the Liability Convention's provisions on damage in space, creates legal complications for removal missions targeting fragments produced by another state's space activities. These complications can be resolved through bilateral arrangements between the launching and removing states, but no general framework exists. Developing such a framework is one of the active issues in current debris policy discussions.

A fourth gap is the limited treatment of cislunar activity in the existing framework. The Outer Space Treaty applies in cislunar space as it does elsewhere, but the technical guidelines developed by the IADC and COPUOS were written with Earth-orbiting activity in mind. The Artemis Accords address some aspects of lunar surface activity but do not comprehensively address cislunar orbital debris. As the Artemis program and subsequent lunar exploration advance, this gap will become more consequential. Chapter 16 examines the cislunar question in more detail.

A fifth gap is the lack of mechanisms to address the cumulative environmental impact of space activity, distinct from individual debris-generating events. The atmospheric impact of reentries discussed in Chapter 6, the astronomical impact of large constellations discussed in Chapter 15, and the broader sustainability questions all involve cumulative effects that no single state's activity produces and that no single state's regulatory framework can address. The development of cumulative-impact frameworks is a growing focus in international debris policy, but no cohesive framework has yet emerged.

A sixth gap, less often discussed but increasingly consequential, is the limited direct accountability of commercial operators under international law. The five core treaties address states, not companies. When a commercial operator based in one state launches from a second state and provides services to customers in a third, the chain of treaty-based accountability spans multiple jurisdictions, each of which must translate the international obligation into domestic regulation and then enforce it. The weakest link in that chain sets the practical floor for accountability. As the commercial sector now accounts for most new launches and an overwhelming share of active satellites, the treaty framework's exclusive focus on state responsibility is an increasingly serious structural limitation. Proposals to adapt international space law to address commercial operators more directly have been made in various forms but have not yet produced treaty-level change.

Prospects for Reform

The prospects for reform of the international space governance framework are mixed. On the one hand, the international community has demonstrated significant capacity to develop non-binding guidelines and norms that, in practice, have shaped state behavior. The IADC mitigation guidelines, the COPUOS LTS Guidelines, and the Artemis Accords are meaningful additions to the framework, even though none has the legal force of a treaty. On the other hand, the international community has not produced new treaties on space governance in more than four decades, and prospects for such treaties remain limited in the current geopolitical environment. The substantive constraints on debris-generating activity are likely to develop primarily through national regulatory frameworks, supported and shaped by international guidelines, rather than through new treaty-level commitments.

Several specific reform proposals are under active discussion. A UN-administered framework for coordinating active debris removal, which would address some of the legal complications discussed above, has been proposed and is under study within COPUOS. Expanding the Liability Convention to address debris-generation activity directly, rather than only damage from such activity, has been proposed but faces substantial diplomatic obstacles. An integrated reporting framework that would extend the Registration Convention to include current operational status, position, and end-of-life plans for active satellites has been proposed and has attracted some support. None of these proposals has yet produced substantive change, but their continued discussion suggests that the international framework is not entirely static.

The role of commercial operators in the international framework is also evolving. The framework's traditional state-to-state structure reflects the era when space activity was conducted exclusively by states and their direct agencies. The growing role of commercial operators, particularly the major constellation operators, has created pressure for institutional adaptations that would enable direct commercial participation in international policy discussions. Various proposals for industry consultation, observer status, or formal advisory roles have been discussed, but none have yet been formally adopted.

Treaty	Adopted	Entered Into Force	Parties (mid-2026)
Outer Space Treaty	1967-01-27	1967-10-10	115
Rescue Agreement	1968-04-22	1968-12-03	99
Liability Convention	1972-03-29	1972-09-01	98
Registration Convention	1975-01-14	1976-09-15	75
Moon Agreement	1979-12-18	1984-07-11	18

Table 7.1. Five Core Space Treaties

Agency	Country / Region	Joined
NASA	United States	1993 (founder)
ESA	Europe	1993 (founder)
JAXA (formerly NASDA)	Japan	1993 (founder)
Roscosmos	Russia	1993 (founder)
CNSA	China	1995
ASI	Italy	1996
BNSC/UKSA	United Kingdom	1996
CNES	France	1996
DLR	Germany	1996
ISRO	India	1996
NSAU	Ukraine	2000
KARI	South Korea	2014
CONAE	Argentina	2017

Table 7.2. IADC Member Agencies

Chapter Summary

The international legal framework governing space activity rests on five core treaties developed during the Cold War and on an extensive body of subsequent guidelines, principles, and norms, primarily developed through the IADC and COPUOS. The framework establishes the basic principles of state responsibility, state liability, peaceful purposes, and free access that have shaped space activity over the past six decades. Its provisions on debris are largely indirect, with the treaty framework's binding obligations supplemented by the non-binding technical and policy guidance of more recent instruments. The Artemis Accords, the LTS Guidelines, and the IADC mitigation guidelines are the most consequential recent additions, and their effective implementation depends largely on individual states' willingness to incorporate this guidance into their national regulatory frameworks. The next chapter examines those national frameworks in which the substantive constraints on debris-generating activity ultimately operate.

Chapter 8. National Regulatory Frameworks

Introduction

The international legal framework examined in Chapter 7 sets out the principles governing space activity but does not directly regulate the operational practices that generate debris. The substantive constraints on debris-generating practices operate primarily through national regulatory frameworks: the licensing rules, design standards, and operational requirements that individual states impose on their nationals' space activities. The framework operates through a chain of accountability that runs from the international level (state responsibility under the Outer Space Treaty) through the national level (domestic licensing rules) to the operational level (specific design and operational requirements imposed on individual missions).

This chapter examines the principal national regulatory frameworks that shape contemporary space activity. The treatment is necessarily selective: more than thirty states maintain some form of national space regulation, and a complete survey is beyond the scope of any single chapter. The focus is on the frameworks of the leading spacefaring states whose practices most directly shape the debris environment: the United States; the European Space Agency member states (treated through national frameworks and the ESA collective framework); the United Kingdom; Japan; China; Russia; and India. Several smaller spacefaring states are also discussed when their practices illustrate specific approaches or when their activities are especially influential in debris policy. The chapter concludes with a discussion of regulatory convergence and divergence, as well as the prospects for greater international harmonization.

The United States Federal Communications Commission

The Federal Communications Commission (FCC) is the principal U.S. regulator of commercial satellite operations and has emerged as one of the most consequential regulators of debris-related practices worldwide. The FCC's authority derives from its general jurisdiction over radio communications and from the requirement that any commercial satellite operator providing communications services in the United States obtain an FCC license to use radio frequencies. Because virtually all commercial satellites rely on radio communications for some aspect of their operation, the FCC license requirement is functionally equivalent to a comprehensive licensing requirement for U.S. commercial satellites and for foreign-licensed satellites that provide services in the United States.

Over the past two decades, the FCC has used its licensing authority to establish robust debris-mitigation requirements. The original framework, adopted in 2004, required license applicants to submit an "orbital debris mitigation plan" addressing specific risk categories: the release of operational debris, accidental explosions, intentional break-ups, end-of-life disposal, and reentry-related casualty risk. The framework was largely procedural; it required disclosure of the operator's mitigation approach but did not impose specific substantive requirements beyond the general expectation of compliance with the IADC guidelines.

The framework was dramatically revised in September 2022 with the adoption of the five-year disposal rule (FCC 22-74). The rule requires that any U.S.-licensed satellite operating in low Earth orbit be removed from orbit within five years of mission completion, replacing the previous twenty-five-year guideline. It applies to satellites disposed of after September 2024 and marks one of the first instances of a national regulator imposing requirements stricter than the international consensus. The substantive justification for the change was that the twenty-five-year disposal guideline had proven inadequate to limit debris accumulation, particularly given the rapid expansion of constellation activity, and that more aggressive disposal was technically achievable for the dominant categories of low Earth orbit satellites.

The five-year rule has had far-reaching practical consequences. Satellite designs intended for U.S. licensing have been modified to include sufficient propulsion for active deorbit at the end of life. Constellation operators have updated their disposal plans to meet the shorter timeline. Foreign satellite operators seeking to provide services in the United States have adapted to U.S. requirements, even when their home jurisdictions impose less stringent standards. The rule has also influenced the international policy debate, with several other regulators considering similar tightening of their disposal requirements.

The FCC's authority extends beyond the disposal rule to other debris-related practices. The agency has imposed conditions on specific licenses requiring enhanced collision-avoidance capabilities, reviewed individual constellation designs for cumulative debris impacts, and begun coordinating with other federal agencies on broader debris issues. The FCC's role has grown rapidly as space activity has become predominantly commercial, and the agency's positions on debris issues are now closely watched by the international space policy community.

The FCC has also begun addressing the risk of casualties from satellite reentries, an area traditionally handled by other federal agencies. The FCC's 2024 adoption of new rules requiring detailed casualty risk analysis for satellite reentries from FCC-licensed missions further extended the agency's debris jurisdiction into the operational practices of constellation operators. The expansion has been controversial, with some industry observers arguing that the FCC is exceeding its statutory authority. In contrast, others argue that consolidated regulation under a single agency provides clearer compliance pathways than regulation fragmented across multiple agencies.

The Federal Aviation Administration's Office of Commercial Space Transportation

The Federal Aviation Administration's Office of Commercial Space Transportation (FAA AST) regulates commercial launch and reentry operations in the United States. The office's authority derives from the Commercial Space Launch Act of 1984, as amended, which requires that any commercial launch from U.S. territory, by U.S. nationals, or carrying U.S. payloads obtain an FAA launch license. The office's regulatory framework focuses on safety, including ground safety during launch operations and the safety of populations and property along the launch trajectory and reentry footprint.

The FAA's debris-related requirements focus primarily on the casualty risk from reentries and launch-phase risks from planned and contingency events. The agency requires detailed analyses of the casualty risk for each licensed launch, including the risk posed by any objects placed in orbit that later reenter Earth's atmosphere. The casualty risk analysis applies to the 1-in-10,000 threshold discussed in Chapter 6 and must demonstrate that the proposed launch and any associated reentries comply with it.

The FAA's framework is distinctive among U.S. agencies because it explicitly addresses the launch-vehicle stages that reach orbit, including requirements for their disposal. It requires that upper stages capable of reaching orbit be designed for either controlled reentry or atmospheric decay within applicable timelines. The framework has been the principal regulatory tool, ensuring that U.S.-licensed launch vehicles do not produce uncontrolled reentries comparable to those of the Long March 5B series. U.S. requirements have consistently been stricter than international guidelines, and U.S. launch vehicles have generally complied with them.

The FAA has also begun to address in-orbit operations of commercial spacecraft, an area that has historically been split among multiple agencies. The 2023 release of new operational rules for commercial space stations and other in-orbit activities extended the FAA's regulatory reach into operational practices that had previously been less clearly regulated. This expansion is part of a broader trend in U.S. space regulation toward consolidation under fewer agencies with broader authority, though the consolidation has been gradual and incomplete.

The National Oceanic and Atmospheric Administration

The National Oceanic and Atmospheric Administration (NOAA), through its Office of Space Commerce, regulates commercial remote sensing activities in the United States. The Office's authority derives from the Land Remote Sensing Policy Act of 1992, as amended, and applies to any commercial operator that provides satellite-based imaging or other remote sensing data. Licensing requirements include national security and foreign policy considerations, as well as technical requirements for spacecraft operations.

NOAA's role in debris policy has been less prominent than that of the FCC or FAA, but it has expanded in recent years. The agency has begun incorporating debris-mitigation requirements into its remote-sensing licenses, drawing on the broader U.S. interagency debris policy framework. It has also been a key participant in federal coordination of debris-related research and has hosted recent atmospheric research on stratospheric metals from spacecraft reentries, as discussed in Chapter 6.

The 2018 establishment of the Office of Space Commerce as a distinct entity within NOAA, with broader responsibilities for space situational awareness and commercial space coordination, has positioned the agency for an expanding role in debris-related policy. The Office has been designated as the lead U.S. civilian agency for providing space situational awareness data to commercial operators, a role previously performed by the U.S. military. The transition from military to civilian provision of conjunction warnings, formally announced in 2021 and implemented over the following years, represents a significant institutional development with implications for both U.S. domestic debris policy and international cooperation on space situational awareness.

Department of Defense and Space Command

The U.S. Department of Defense, primarily through Space Command and the U.S. Space Force, plays a central role in debris-related operations, even though its regulatory authority is more limited than that of the civilian agencies. Space Command operates the Space Surveillance Network, the global network of radar and optical sensors that maintains the principal U.S. catalog of orbital objects. The Eighteenth Space Defense Squadron, based at Vandenberg Space Force Base in California, manages the day-to-day operations of the Space Surveillance Network and provides conjunction warnings to satellite operators worldwide.

DOD's debris-related operations are notable for several characteristics. First, they extend beyond U.S. national satellites, providing services to operators of all nationalities through cooperative arrangements. The Combined Space Operations Center, established at Vandenberg in 2009, coordinates these international services. Second, the DOD's operations involve extensive classified activity that is not visible to the international policy community, including specific tracking capabilities, conjunction analyses for sensitive satellites, and military activities that are not formally part of the public catalog. The relationship between the unclassified and classified components of the U.S. tracking infrastructure has been a recurring topic of discussion within the international policy community.

The transition of conjunction services from the DOD to NOAA, beginning in 2021 and continuing through the mid-2020s, is an institutional development with significant implications. The transition reflects a U.S. policy decision that civilian provision of space situational awareness is preferable for commercial operators, both for institutional efficiency and for international cooperation. The transition has been gradual, with NOAA's Office of Space Commerce building the capability to deliver SSA services that the DOD has historically provided. Whether the transition delivers equivalent or better service to commercial operators while maintaining the security-relevant aspects of the DOD's tracking infrastructure will be a critical test of U.S. space governance over the coming decade.

The European Space Agency and Member State Frameworks

The European Space Agency (ESA) is an intergovernmental organization comprising 22 member states. ESA does not have direct regulatory authority over commercial space activity in member states; that authority rests with each member state's national regulators. ESA's role is operational and coordinating: the agency develops and implements specific space programs, coordinates among member-state agencies, and develops technical standards and guidance that member states adopt to varying degrees within their national regulatory frameworks.

ESA's internal debris-mitigation framework is among the world's most stringent. The agency's Space Debris Mitigation Policy, most recently revised in 2023, requires ESA missions to meet strict debris-related requirements that exceed international guidelines in several areas.

The framework mandates post-mission disposal within five years for most low Earth orbit missions, controlled reentry with a 1-in-100,000 risk of casualties for certain mission categories, comprehensive passivation of upper stages, and specific design requirements to prevent fragmentation. The framework is binding on ESA missions and has substantially shaped the design of European spacecraft and launch vehicles over the past decade.

National regulatory frameworks among ESA member states vary widely. France has one of the most developed national space laws (the Law on Space Operations of 2008), which establishes a comprehensive licensing framework for French space activities and incorporates debris-mitigation requirements that closely align with the IADC framework. Germany's space activities are regulated under a more fragmented framework that draws on telecommunications and transportation law, as well as specific space-related statutes. The United Kingdom, no longer an ESA member at the regulatory level since Brexit, has developed an independent national framework closely aligned with ESA practices.

Italy operates one of the more developed European national frameworks, with the Italian Space Agency exercising both implementation and regulatory functions for Italian space activities. The Netherlands, Belgium, Sweden, and several other smaller ESA member states have national frameworks that delegate primary regulatory authority to their national space agencies. Spain has developed a comprehensive national framework that incorporates debris-mitigation requirements aligned with European norms.

The European Union has increasingly become a relevant actor in space policy, despite its limited direct regulatory authority. The EU's Galileo navigation satellite program, the Copernicus Earth observation program, and the EU Space Surveillance and Tracking framework have brought the Union into operational space activities that complement the ESA framework. The EU Space Program Regulation of 2021 provides the legal basis for the Union's space activities and includes debris mitigation provisions aligned with ESA and IADC standards. The Union has not yet developed a comprehensive regulatory framework for commercial space activity equivalent to those of major member states, but discussions on such a framework are ongoing.

The United Kingdom Space Agency

The United Kingdom's space regulatory framework has evolved significantly since the Space Industry Act of 2018, which established a comprehensive licensing framework for UK space activities. The UK Space Agency, established in 2010, performs both implementation and regulatory functions, including licensing commercial space activities under the Space Industry Act and its associated regulations.

The UK framework includes specific debris-mitigation requirements that have been progressively tightened in recent years. The 2021 revision of the UK regulations established a five-year post-mission disposal requirement for low Earth orbit missions, similar to the U.S. FCC rule but adopted before the formal U.S. rule in 2022. The UK framework also includes requirements for passivation, controlled reentry where applicable, and casualty risk analysis. These requirements apply to all UK-licensed missions, including the growing number of commercial operators based in the United Kingdom.

The UK has also developed a robust space situational awareness capability through the National Space Operations Centre. This joint civilian-military facility monitors UK satellites and provides conjunction screening services. Although smaller than those of the U.S. or ESA, the capability gives the United Kingdom independent operational capacity for the growing number of UK-licensed satellites. The UK's approach has emphasized integration with international SSA networks, including extensive cooperation with the U.S. Space Surveillance Network, while developing the indigenous capability needed for sovereign space operations.

The UK's post-Brexit position outside the ESA framework has created both challenges and opportunities for UK space policy. The challenges include the loss of direct participation in certain ESA programs and the need to develop independent capabilities that previously relied on ESA cooperation. The opportunities include the freedom to develop regulatory frameworks tailored to the UK industry, the ability to establish bilateral agreements with non-European partners, and the development of distinctive UK capabilities that differentiate the UK industry within the broader European space sector.

The Japanese Aerospace Exploration Agency

The Japan Aerospace Exploration Agency (JAXA), established in 2003 through the consolidation of three previous agencies, is Japan's principal space agency and performs both operational and regulatory functions. The Japanese space regulatory framework is based on the Space Activities Act of 2008 and subsequent revisions, which set licensing requirements for Japanese space activities and incorporate debris-mitigation requirements aligned with the IADC framework.

The Japanese framework is notable for several characteristics. It is among the most rigorous in applying IADC standards, with consistent compliance across its national space programs. It has been progressively tightened to incorporate developments in international guidelines. It has also emphasized integrating debris-mitigation considerations into the broader strategic framework for Japanese space activities, including specific requirements for compliance reporting and verification.

JAXA's operational activities include significant contributions to debris research and mitigation. The agency operates Earth observation, navigation, and scientific research satellites that comply with Japan's mitigation framework. The agency has developed and demonstrated several debris-related technologies, including electrodynamic tether concepts for active deorbit. The Japanese commercial sector has been a major developer of active debris removal technology, with Astroscale, a Japanese-incorporated commercial debris-removal company, conducting the first operational demonstration of debris capture and disposal techniques.

The Japanese framework's effectiveness has been demonstrated in handling specific incidents, including the 2016 Hitomi breakup discussed in Chapter 5. The post-event analysis identified procedural improvements that were later incorporated into JAXA practices. The Japanese framework's transparency in addressing failures has been markedly better than that of some other leading spacefaring states, and the lessons learned from Hitomi have contributed to the broader international understanding of software-related fragmentation risks.

The China National Space Administration

The China National Space Administration (CNSA) is the principal Chinese civilian space agency, while the People's Liberation Army Strategic Support Force (PLASSF) manages military space activities. The Chinese space regulatory framework is less transparent than that of leading Western spacefaring states, and much of operational practice remains undocumented. What is publicly known about the framework comes primarily from official Chinese policy statements, observed operational practices, and international cooperation activities.

The Chinese framework has evolved substantially over the past two decades. China is a member of the IADC and has formally subscribed to the committee's guidelines. China participated in the development of the COPUOS LTS Guidelines and has formally endorsed them. The 2016 Chinese white paper on space activities included specific commitments to debris mitigation, and subsequent policy documents have reinforced them. In practice, however, China's implementation of the international guidelines has been uneven. The 2007 Fengyun-1C anti-satellite test predates much of China's current commitment but remains the largest single contributor to the contemporary debris environment. The Long March 5B uncontrolled reentries discussed in Chapter 6 have continued during the period in which China has formally subscribed to the international guidelines.

The Chinese commercial space sector has expanded rapidly in recent years, with multiple commercial launch vehicle developers and satellite operators emerging since approximately 2015. The regulatory framework for commercial activity in China has evolved alongside the sector but remains less transparent than Western frameworks. Commercial operators are typically required to obtain government approval for their activities. They are subject to general regulatory oversight, but the specific requirements, particularly those related to debris mitigation, are less publicly documented than those in the United States or Europe.

The Chinese satellite constellation programs, including Guowang and Qianfan, have been deployed under regulatory frameworks that include some debris-mitigation requirements but have not undergone detailed international examination. The constellations are intended to provide global coverage and have been deployed across multiple altitude shells, with implications for the debris environment that depend heavily on operators' operational practices. Whether the constellations achieve the disposal compliance levels assumed in the international debris analysis has not been independently verified.

Guowang, operated by the state-backed Shanghai Spacecom Satellite Technology, has filed ITU paperwork for up to 12,992 satellites across multiple orbital shells, with some proposed at altitudes up to 1,145 kilometers. Qianfan, also known as Thousand Sails, is being developed by Shanghai Yuanxin Satellite Technology and has proposed orbital shells between 600 and 1,000 kilometers. Neither operator has published detailed disposal compliance plans comparable to those of SpaceX, OneWeb, or Amazon Leo. The ITU filings establish frequency coordination rights but do not, in themselves, impose debris-mitigation requirements. Whether either constellation will achieve disposal performance consistent with the five-year rule adopted by leading Western regulators depends on regulatory requirements that the Chinese licensing framework has not yet publicly specified and on operational execution that has not yet been demonstrated at scale.

The strategic dimension of China's space activity remains a central consideration in the international debris policy discussion. China has demonstrated anti-satellite capability multiple times, though it has not conducted destructive testing since 2007. The retention of anti-satellite capability, combined with the strategic competition between China and the United States, creates persistent uncertainty about whether China would conduct further destructive testing in a future strategic crisis. The international community's capacity to prevent such testing depends largely on the strength of the international norm against destructive testing, which Chapter 5 examines in more detail.

The Russian Federation

Russia's space regulatory framework has evolved through several major transitions over the past three decades. The Soviet space program operated under a highly classified framework and through state agencies. The post-Soviet transition produced a fragmented framework that incorporated elements of Soviet practice while adapting to the new political and economic context. The Roscosmos State Corporation, established in its current form in 2015, is the principal Russian space agency and exercises both operational and regulatory functions.

The Russian framework includes formal adherence to the IADC guidelines and the COPUOS LTS Guidelines, and Russia participates in international policy discussions on space debris. In practice, Russian implementation has been inconsistent. The legacy of unpassivated Soviet-era upper stages, which continue to explode at intervals, as discussed in Chapter 5, remains a persistent source of new debris that Russian regulatory action has not fully addressed. The 2021 Cosmos 1408 anti-satellite test, conducted by the Russian Federation, was the most consequential debris-generating event in recent years.

The Russian commercial space sector is smaller than those of the United States, China, or Europe, and its regulatory framework for commercial activity is correspondingly less developed. Russian commercial operators are generally subject to government oversight through Roscosmos, but the debris-mitigation requirements imposed on commercial activity are less publicly documented than in Western frameworks.

The deterioration of Russia's relations with Western spacefaring states since 2014, particularly after the 2022 invasion of Ukraine, has affected Russia's role in international debris policy. Operational cooperation on the International Space Station has continued, but broader cooperation has declined significantly. The 2021 Cosmos 1408 test, conducted amid escalating tensions preceding the invasion of Ukraine, illustrated the Russian government's willingness to engage in debris-generating activities even at a high international diplomatic cost. Whether Russian behavior on debris policy will improve as broader relations stabilize, or remain a source of persistent friction, depends heavily on developments in the broader strategic environment.

The Indian Space Research Organization

The Indian Space Research Organization (ISRO) is India's principal space agency and performs both operational and regulatory functions. India's space regulatory framework has been formalized by the 2023 Indian Space Policy, which established a comprehensive framework for government and commercial space activities. The framework includes debris-mitigation requirements aligned with the IADC framework, although specific implementation details have not been fully documented in publicly available sources.

India's role in debris policy has expanded significantly in recent years. The 2019 Mission Shakti anti-satellite test, discussed in Chapter 5, established India as a state with demonstrated anti-satellite capability. The choice of a low altitude for the test reflected an awareness of the consequences of debris, distinguishing the Indian event from the higher-altitude Chinese test of 2007. India has not conducted a follow-on test, although Indian space officials have stated that the capability remains available.

ISRO's operational programs include extensive Earth observation, navigation, and scientific missions, all of which are subject to India's regulatory framework. The Indian commercial space sector has expanded rapidly since the 2020 reforms that opened commercial space activities to private participation, with multiple commercial launch vehicle developers and satellite operators emerging in the years that followed. The regulatory framework for commercial activity has been progressively developed, with the 2023 policy and subsequent guidance documents establishing the operational framework.

India's position on debris policy reflects its broader strategic posture as a nonaligned state seeking to balance its relationships with the major powers. India has been an active participant in COPUOS and IADC discussions and has subscribed to the international guidelines. Still, India has retained strategic autonomy in specific areas, including maintaining an anti-satellite capability. The growth of India's commercial space sector, combined with the country's strategic position, makes India a state whose regulatory choices will strongly influence the international debris environment in the coming years.

Smaller Spacefaring States

A growing number of smaller spacefaring states have developed national regulatory frameworks for their space activities. The selection below is necessarily incomplete but illustrates the range of approaches.

Luxembourg has developed one of the more progressive national frameworks for commercial space activity, including the 2017 law on the exploration and use of space resources, which provides legal certainty for such activities within Luxembourg's jurisdiction. The framework includes debris-mitigation requirements aligned with IADC and ESA standards. Luxembourg's framework has attracted commercial operators seeking a clear regulatory environment, contributing to the country's growing role as a space industry hub.

The United Arab Emirates has developed a comprehensive national space framework through the UAE Space Agency, established in 2014. The framework includes debris-mitigation requirements aligned with international standards and applicable to both government and commercial space activities. The UAE has been an active participant in international space policy discussions and has subscribed to the LTS Guidelines and the Artemis Accords.

Australia, through the Australian Space Agency established in 2018, has developed a national framework that includes debris-mitigation requirements aligned with international standards. The framework places particular emphasis on space situational awareness, with significant investment in tracking capabilities that contribute to the international SSA network.

The Republic of Korea has a national framework administered by the Korea Aerospace Administration, which was reorganized in 2024. The framework includes debris-mitigation requirements consistent with international standards and applies to both government and commercial activities. Korea has rapidly expanded its space capabilities in recent years, developing indigenous launch vehicles and a growing satellite constellation.

Brazil, through the Brazilian Space Agency, has developed a national framework that includes debris-mitigation requirements, but it has been applied less rigorously than in the major spacefaring states. Brazil's space activities have been more limited in scope than those of larger states, and the regulatory framework has accordingly received less attention.

New Zealand has developed a distinctive national framework that focuses primarily on commercial launch activities, particularly those conducted by Rocket Lab from the Mahia Peninsula launch site. The framework includes debris-mitigation requirements that have been consistently applied to Rocket Lab operations.

Several African states have developed national space frameworks of varying degrees of formality. The African Space Agency, established in

2018 with its headquarters in Cairo, has begun coordinating space policy among African states but has not yet produced a comprehensive continental regulatory framework.

Regulatory Convergence and Divergence

National regulatory frameworks reveal both meaningful convergence and substantial divergence. Convergence has been driven primarily by the IADC and COPUOS guidelines, which have provided a common technical baseline adopted by most leading regulators. The IADC's twenty-five-year disposal rule served as the international standard for two decades and was incorporated into national frameworks across multiple jurisdictions. The COPUOS LTS Guidelines have similarly provided a common framework that most leading regulators adopt in some form.

Several factors have driven divergence. National strategic considerations, particularly those related to military and security space activities, have led to regulatory carve-outs in many jurisdictions that limit the application of debris-mitigation requirements to specific activities. National industrial policy considerations have led to different approaches to commercial space regulation, with some states adopting more permissive measures to attract industry and others adopting more stringent ones based on environmental or safety concerns. National technical capacity has affected the implementation of common standards, with states that have stronger technical infrastructure able to apply more sophisticated requirements than those with less developed capacity.

The most consequential current trend in national regulation is the tightening of disposal requirements in major jurisdictions. The U.S. five-year rule of 2022, the UK five-year rule of 2021, and the ESA five-year requirement for most missions all represent markedly tighter standards than the previous twenty-five-year baseline. The trend is likely to continue, with several other regulators considering similar tightening. However, the pace of tightening has not been uniform. Some leading spacefaring states have not adopted equivalent, tighter standards, and the divergence between the leaders and the laggards has become increasingly pronounced.

A second important trend is the integration of debris considerations into broader regulatory frameworks. While debris was historically a specialized topic addressed by specific guidelines, it has increasingly become part of comprehensive regulatory frameworks that govern all aspects of space activity. Both the growing recognition of debris as a central regulatory concern and the practical reality that debris-mitigation requirements affect spacecraft design and operation in ways that interact with other regulatory requirements have driven this integration.

A third trend is the growing role of commercial considerations in regulatory design. The rising share of space activity conducted by commercial operators has put pressure on regulatory frameworks to be predictable, transparent, and proportionate to the risks they address. The U.S., UK, and ESA frameworks have all evolved in response to this pressure, with increasingly detailed published guidance and more structured processes for license applications and compliance verification. Developing commercial-friendly regulatory frameworks has been a principal way for jurisdictions to compete for commercial space activity.

The Role of Industry Standards

Beyond formal regulatory frameworks, industry standards developed by bodies such as the International Organization for Standardization (ISO) play an important role in shaping debris-related practices. ISO 24113, the standard for space systems debris mitigation, provides a detailed technical framework for meeting debris-mitigation requirements. The standard is closely aligned with the IADC guidelines and has been adopted as the basis for compliance verification across multiple regulatory frameworks.

ISO standards have several characteristics that complement the formal regulatory framework. They are developed through the participation of technical experts rather than through diplomatic negotiation, producing more detailed technical guidance than is typically possible in regulatory documents. They are updated more frequently than regulatory frameworks, allowing them to incorporate new technical understanding more quickly. They provide a common technical reference that operators across multiple jurisdictions can use, reducing compliance costs for those subject to multiple regulators.

The Space Sustainability Rating program, originally initiated by the World Economic Forum and now operated by the École Polytechnique

Fédérale de Lausanne in Switzerland, offers a complementary mechanism for evaluating and disclosing the sustainability characteristics of space missions. The program assigns ratings based on criteria such as debris-mitigation practices, conjunction-avoidance capabilities, operational transparency, and adherence to international guidelines. The ratings are voluntary, yet they have begun to influence commercial decisions, including insurance pricing and customer selection.

Several other industries have emerged in recent years to address specific aspects of debris-related practices. The Space Safety Coalition, an industry-led initiative comprising major satellite operators, has developed best practices for collision avoidance and constellation operations. The Consortium for Execution of Rendezvous and Servicing Operations (CONFERS) has developed standards for in-orbit servicing operations, which are relevant to active debris removal. These initiatives complement formal regulatory frameworks by providing detailed operational guidance and promoting the adoption of best practices across the industry.

Compliance Verification and Enforcement

A critical question for any regulatory framework is how compliance is verified and enforced. The frameworks discussed in this chapter rely on a combination of pre-launch review, post-launch monitoring, and end-of-life verification, with considerable variation across jurisdictions in the rigor of these measures.

Pre-launch review typically requires the operator to submit documentation demonstrating compliance with applicable requirements for the proposed mission, including debris mitigation plans, casualty risk analyses, and disposal plans. The review can range from procedural verification that required documentation has been submitted to detailed technical analysis of the operator's claims. The U.S. and ESA frameworks include extensive technical review, whereas some other frameworks rely more heavily on operator self-certification.

Post-launch monitoring involves tracking actual operational practices against the commitments made during the pre-launch review. Monitoring is typically less intensive than the pre-launch review and relies heavily on the operator's voluntary reporting of operational events. Conjunction-avoidance maneuvers, fragmentation events, and other debris-relevant operational decisions are typically reported to the regulator, but verification of these reports is limited.

End-of-life verification is the primary mechanism for ensuring that disposal commitments are met. The verification typically requires the operator to demonstrate that the satellite has been removed from operational orbits within the required timeline. Compliance is generally good for satellites operating at altitudes where natural decay assists with disposal, more variable for those operating at altitudes where active deorbit is required, and most variable for satellites that have failed or lost contact before reaching the end of life.

Enforcement mechanisms are limited across most frameworks. The principal sanction is the revocation or non-renewal of licenses, which affects the operator's future activities but does not address the specific compliance failure. Financial penalties are rare, and criminal sanctions are essentially nonexistent. The limited enforcement reflects both the limited investigative capacity of most regulators and the difficulty of attributing specific debris-generating events to specific operator decisions. As a result, compliance depends heavily on operator commitment rather than on regulatory enforcement, and the framework's effectiveness largely depends on the alignment of operator interests with regulatory objectives.

The Path Forward for National Frameworks

The evolution of national regulatory frameworks is likely to continue along several lines over the coming decade. Disposal requirements are likely to tighten, with progressively shorter timelines and more stringent compliance verification. The integration of debris considerations into broader regulatory frameworks is likely to deepen, with more comprehensive treatment of operational practices that affect debris generation. The development of commercial-friendly regulatory designs is likely to continue, with more transparent and predictable processes that reduce compliance costs while maintaining substantive protection.

Several specific reforms are under active discussion in major jurisdictions. The U.S. is considering further consolidating space regulation under fewer agencies with broader authority. Several European jurisdictions are developing national frameworks for active debris removal services to address some of the legal complications discussed in

Chapter 7. Japan is considering further integrating commercial debris-removal capabilities into its national framework. The United Arab Emirates and several other emerging spacefaring states are considering adopting more comprehensive frameworks to position themselves as preferred jurisdictions for specific categories of space activity.

The international harmonization of national frameworks remains a significant challenge. Despite common reference points provided by the IADC and COPUOS guidelines, variations in national practices impose transaction costs on operators who must satisfy multiple regulators. Various harmonization efforts have been proposed, including bilateral and multilateral mutual recognition arrangements that would allow compliance with one jurisdiction's requirements to satisfy the requirements of others. Progress on harmonization has been slow, as differences in national strategic and industrial considerations continue to produce divergence even as the technical baseline converges.

The growing role of national regulatory frameworks in shaping the orbital debris environment makes the trajectories of these frameworks central determinants of long-term sustainability. The international guidelines provide the technical baseline, but the substantive constraints that shape operational practice are set at the national level. The pace at which national frameworks tighten their requirements, the consistency of compliance verification across jurisdictions, and the willingness of national regulators to address the specific issues posed by large constellations and active debris removal will collectively determine whether the international debris policy framework yields meaningful improvement in the orbital environment.

Jurisdiction	LEO Disposal Timeline	Year Adopted	Effective Date
United States (FCC)	5 years	2022	2024 (phased)
United Kingdom	5 years	2021	2022
ESA (member missions)	5 years (most missions)	2023	2024

Jurisdiction	LEO Disposal Timeline	Year Adopted	Effective Date
France (CNES)	25 years (under review)	2008	Active
Japan	25 years	2018	Active
IADC (international)	25 years	2002 (rev. 2021)	Voluntary baseline
ISO 24113	25 years	2019	Voluntary standard

Table 8.1. Disposal Timeline Requirements by Jurisdiction (mid-2026)

Country	National Agency	Principal Regulator(s)
United States	NASA	FCC, FAA, DOD, NOAA Office of Space Commerce
Russia	Roscosmos	Roscosmos State Corporation
China	CNSA	State Administration of Science, Technology, and Industry
Japan	JAXA	Cabinet Office; Ministry of Education, Culture, Sports, Science and Technology
India	ISRO	Department of Space; IN-SPACe
Europe (multinational)	ESA	National authorities; EU Space Programme
United Kingdom	UKSA	Civil Aviation Authority (Space Industry Act)
France	CNES	Ministry of Higher Education, Research and Innovation
Germany	DLR	BMWi (Federal Ministry of Economic Affairs)

Table 8.2. National Space Agencies and Principal Regulatory Authorities

Chapter Summary

The international legal framework sets the principles governing space activity, but substantive constraints on debris-generating practices are enforced primarily through national regulatory frameworks. The U.S. framework, centered on the FCC, FAA, NOAA, and DOD, has emerged as one of the most consequential in the world, particularly through the FCC's adoption of the five-year disposal rule in 2022. The European framework, operating through ESA's internal requirements and the national frameworks of member states, has been among the most stringent and has progressively tightened over the past decade. The UK, Japanese, Indian, Chinese, Russian, and other national frameworks vary in stringency and transparency, with notable divergence in the specific requirements imposed on operators. Across leading frameworks, the trend has been toward more stringent disposal requirements, more comprehensive integration of debris considerations into broader regulatory frameworks, and more commercially friendly process design that maintains substantive protection while reducing compliance costs. The variation in national practice remains substantial, and the international harmonization of frameworks, through bilateral arrangements, mutual recognition, or coordinated tightening of standards, will be one of the principal challenges of debris policy over the coming decade. The chapters that follow examine the technical infrastructure, operational practices, and economic systems within which these regulatory frameworks operate.

Chapter 9. Tracking, Space Situational Awareness, and Space Traffic Management

Introduction

The orbital debris environment cannot be managed without measurement. Every aspect of debris policy, from designing mitigation rules to planning removal missions to the daily operational decisions of satellite operators, rests on the empirical foundation provided by the global infrastructure for tracking objects in orbit. The infrastructure has evolved over six decades from a handful of military radars built in the late 1950s to a globally distributed network of sensors, data systems, and analytical capabilities that processes hundreds of thousands of operations per day. This chapter examines the technical, institutional, and operational dimensions of that infrastructure: the sensors that produce raw observational data, the data systems that catalog and propagate orbital elements, the conjunction screening processes that translate the data into operational warnings, and the broader concept of space traffic management that has emerged as the framework for coordinating all of these activities.

The chapter also examines the emerging question of how the existing infrastructure must adapt to the contemporary environment. The traditional infrastructure was designed for a population of a few thousand cataloged objects, operated predominantly by national programs, with manual processes for warning distribution and limited analytical capacity. The contemporary environment includes more than forty thousand cataloged objects, eleven thousand active satellites, and a growth trajectory that points toward populations several times larger within the next decade. Whether the infrastructure can scale to meet this challenge and how the institutional arrangements governing it should adapt are among the most pressing questions in contemporary space policy.

The Origins of Space Surveillance

The U.S. space surveillance infrastructure dates to the late 1950s, when the launch of Sputnik 1 made tracking artificial satellites a national security concern. The original infrastructure consisted of a small number of military radars, primarily intended for missile warning, that were adapted to track satellites as a secondary mission. The Naval Space Surveillance System, established in 1958, used a network of transmitter and receiver stations across the southern United States to detect satellites passing through a fan-shaped beam sweeping the sky. The Air Force Spacetrack system, established the same year, used radars at Eglin Air Force Base in Florida and other sites to provide complementary tracking.

The infrastructure expanded steadily through the 1960s and 1970s as the population of orbital objects grew, and the strategic importance of satellite reconnaissance and communications increased. New radars were added at Thule Air Base in Greenland, Clear Air Force Station in Alaska, Cavalier Air Force Station in North Dakota, and several other sites. Optical sensors were established at Maui, Hawaii; at Mount John in New Zealand; and at several other locations, providing complementary observational capabilities at higher altitudes where radar coverage was less effective. The combined network was designated the Space Detection and Tracking System and was operated by what eventually became the United States Space Surveillance Network.

The Soviet Union developed an equivalent capability through its own network of radars and optical sensors. The Soviet system, operated by the Space Surveillance Service of the Soviet armed forces, was less publicly documented than the American system but followed similar principles and produced an independent catalog of orbital objects. The Russian Federation inherited the Soviet space surveillance system after 1991 and has continued to operate it, though with reduced capability relative to the Soviet-era system and heavy reliance on Russian-developed sensor technology.

Events in recent decades have profoundly shaped the development of contemporary infrastructure. The 2007 Fengyun-1C anti-satellite test showed that existing tracking capabilities were inadequate to characterize a major fragmentation event in real time; the U.S. Space Surveillance Network required weeks to catalog the resulting fragments fully. The 2009 Iridium-Cosmos collision showed that existing conjunction screening processes were inadequate to prevent catastrophic accidents; the predicted miss distance was dramatically wrong. The deployment of large constellations beginning in the late 2010s has strained the catalog and screening infrastructure, driving rapid expansion and modernization. Each of these episodes has prompted institutional responses that have shaped the contemporary infrastructure.

Sensor Types and Capabilities

The global tracking infrastructure relies on three principal sensor types, each with distinct capabilities and limitations.

Radar sensors detect orbital objects by transmitting radio-frequency energy and measuring the reflected signal. A radar's detection range depends on transmitted power, antenna size, operating frequency, and the target's radar cross-section. Modern phased-array radars, with thousands of individually steerable antenna elements, can track multiple objects simultaneously and provide high update rates. The U.S. Space Surveillance Network's principal radars include the Eglin AN/FPS-85 phased-array radar in Florida, which can track objects at altitudes up to several thousand kilometers; the Cobra Dane radar at Shemya, Alaska, which provides tracking coverage for the northern polar region; and the Globus II radar at Vardø, Norway, which provides coverage for the high-latitude portions of low Earth orbit. The Space Fence, an S-band phased-array radar at Kwajalein Atoll in the Marshall Islands that became operational in 2020, considerably expanded the U.S. capability to track small objects in low Earth orbit, detecting objects as small as approximately 5 centimeters (2 inches) at altitudes up to 1,500 kilometers (932 miles).

Optical sensors detect orbital objects by observing reflected sunlight. They are most effective for higher-altitude objects, where radar performance declines sharply, and are essential for tracking the geostationary belt and other high-altitude regions. The U.S. Ground-Based Electro-Optical Deep Space Surveillance system, with telescopes at Socorro, New Mexico; Maui, Hawaii; and Diego Garcia in the British Indian Ocean Territory, provides the principal U.S. capability for high-altitude tracking. ESA operates similar capabilities through telescopes in Spain and Chile. Additional optical telescopes operated by national and academic programs contribute to the global tracking effort.

Optical sensors are limited by the need for darkness at the sensor location and sunlight on the target, which restricts observation opportunities to the twilight periods at the sensor location. This constraint is partially addressed by deploying optical sensors at multiple longitudes, ensuring that at least one sensor is always in the appropriate observation geometry. Space-based optical sensors address this limitation more directly by avoiding the atmospheric and geometric constraints of ground-based observations. The U.S. operates the Space-Based Surveillance System, with sensors in low Earth orbit that provide additional coverage for high-altitude objects. ESA's Space Surveillance and Tracking Programme has explored similar capabilities. Commercial operators, including LeoLabs, NorthStar Earth and Space, and others, are developing space-based optical sensors to augment the existing capability.

Laser ranging provides a third tracking modality that complements radar and optical observation. Satellite laser ranging stations transmit short laser pulses to retroreflectors on cooperating satellites and measure the round-trip time with extreme precision. The technique delivers millimeter-level position accuracy for cooperating satellites and is essential for calibrating the broader tracking infrastructure. The International Laser Ranging Service coordinates a global network of laser ranging stations that provide observational data for hundreds of cooperating satellites. The technique cannot be applied to non-cooperating objects (which lack retroreflectors), so its direct contribution to debris tracking is limited. Still, the high-precision orbital determinations it provides for reference satellites are essential for validating the lower-precision techniques used for general space surveillance.

Each sensor type has distinct limitations that constrain the overall capability of the tracking infrastructure. Radars are most effective at low altitudes and degrade in performance with increasing range; detecting small objects at geostationary altitudes requires impractically large radars. Optical sensors are most effective at high altitudes and degrade in performance for low-altitude objects, whose rapid angular motion limits exposure time; they also depend on weather and viewing geometry. Laser ranging requires cooperating targets and cannot be applied to debris. The complementary use of multiple sensor types is essential to provide reliable coverage across the full range of altitudes and object sizes relevant to debris tracking.

The Catalog of Tracked Objects

The primary product of the global tracking infrastructure is the catalog of objects in orbit. The U.S. catalog, maintained by the Eighteenth Space Defense Squadron at Vandenberg Space Force Base, contains approximately 47,000 objects as of mid-2026, including about 11,000 active satellites and the remainder debris fragments, defunct satellites, and rocket bodies. The catalog is the largest and most comprehensive in the world and serves as the principal reference for international debris policy discussions. This figure exceeds the roughly 40,000 tracked objects cited in Chapter 1, which is drawn from the European Space Agency's annual Space Environment Report. The difference reflects the two organizations' distinct sensor networks, detection thresholds, and counting conventions rather than any disagreement about the underlying environment. The U.S. total is generally higher because the Space Surveillance Network draws on a broader set of sensors and catalogs a larger share of small and uncorrelated objects. At the same time, the ESA figure reflects its own accounting of the tracked population. The Russian catalog, maintained by the Russian space surveillance service, is smaller and less publicly documented but provides an independent reference. China operates its own catalog, which is also less publicly documented than the U.S. catalog.

Each cataloged object is assigned a unique identifier and associated with a set of orbital elements that describe its motion. These elements are typically expressed in the two-line element (TLE) format developed for the U.S. catalog. The TLE includes the orbital period, eccentricity, inclination, right ascension of the ascending node, argument of perigee, mean anomaly, and the time of the elements' validity. The TLE format is compact and human-readable but has limited precision for some applications; modern operational practice often uses higher-precision formats for specific applications, even when the public catalog continues to use TLEs.

The catalog is updated continuously as new observations are processed. Each cataloged object is observed multiple times per day by the sensor network, and its orbital elements are refined using these observations. Objects that have not been observed for an extended period are flagged as potentially decayed or lost; removal from the catalog typically requires an extended absence of observations consistent with reentry. New objects detected by the sensors are evaluated to determine whether they correspond to known launches (in which case they are associated with the launch event) or to fragmentation events (in which case they are associated with the parent object). An assignment can be complicated by the simultaneous appearance of large numbers of fragments after major events.

The catalog's accuracy and completeness depend on sensor performance and on the analytical processes that combine sensor observations to determine orbits. Accuracy is generally good for large, regularly observed objects, with position uncertainties ranging from tens to hundreds of meters (yards) at the time of observation and growing to kilometers (miles) over propagation periods of days. Accuracy degrades for smaller, less frequently observed objects and for longer propagation periods. Accuracy is generally inadequate for fine-grained operational decisions without additional data, and operators requiring high-precision orbital information typically maintain their own tracking capability or subscribe to commercial data services.

The catalog's completeness, the fraction of all objects in orbit that are represented in the catalog, depends on the size threshold at which sensors can reliably detect objects. The U.S. catalog includes essentially all objects larger than approximately 10 centimeters (3.9 inches) in low Earth orbit and approximately 1 meter (3.3 feet) in geostationary orbit. Smaller objects exist in much larger numbers but are not individually tracked; their contribution to the broader debris environment is estimated through statistical analysis of the observed background and impact measurements on the surfaces of returned spacecraft. The cataloging threshold continues to be lowered as new sensors enter service; the Space Fence at Kwajalein has expanded the population of small objects reliably observed in low Earth orbit, and similar expansions are expected as additional sensors are commissioned.

Conjunction Screening and Warning

The primary operational use of the tracking infrastructure is conjunction screening: systematically identifying close approaches between cataloged objects and issuing warnings to operators of objects at risk. The screening process involves propagating the orbits of all cataloged objects forward in time, identifying pairs that come within a defined screening volume, and computing the collision probability for each conjunction. Operators of active satellites receive warnings about predicted conjunctions and are responsible for deciding whether to perform avoidance maneuvers.

The U.S. Combined Space Operations Center, established in 2009 following the Iridium-Cosmos collision, performs conjunction screening for active satellites worldwide. The center provides Conjunction Data Messages (CDMs) to satellite operators when predicted close approaches meet specified criteria. CDMs include the predicted positions of the two objects at closest approach, their position uncertainties, and the computed probability of collision. The criteria for issuing CDMs have evolved over time and now include both miss-distance thresholds (typically a few kilometers) and collision-probability thresholds, commonly set below 1 in 100,000 to maintain operational safety. This threshold corresponds to an estimated collision probability of 0.001% during the predicted close approach.

The probability of collision is the primary operational metric for assessing conjunction risk. It accounts for uncertainties in the predicted positions of the two objects, their combined cross-sectional area, and the geometry of the encounter. A high probability of collision warrants a maneuver decision; a low probability does not. The threshold for a maneuver depends on the operator and the spacecraft. Crewed spacecraft typically use thresholds around 1 in 100,000; any conjunction with a probability above this threshold warrants action. Commercial constellation satellites typically use thresholds around 1 in 10,000, recognizing that the operational disruption from frequent maneuvers must be balanced against collision risk. Some smaller commercial satellites use looser thresholds based on the cost of maneuvering compared to the cost of replacing the satellite.

The screening process must address several technical challenges. Positional uncertainties of cataloged objects can be substantial, especially for objects observed infrequently or for predictions extended over longer periods. The collision probability can be sensitive to assumptions about the objects' cross-sectional areas, which are often poorly known for debris fragments. The screening process must balance the competing demands of comprehensive coverage (catching all potentially dangerous conjunctions) with a manageable workload (avoiding operator overload from low-probability warnings). The thresholds for issuing warnings, the screening volumes used, and the probability calculations applied have all evolved as the operational community has developed more refined approaches.

The dilution problem, mentioned in Chapter 4, warrants repeated emphasis here. When position uncertainties are large, the probability of collision can be artificially low even when the nominal miss distance is small. This occurs because the probability density at any point decreases as the uncertainty distribution spreads. Operators relying solely on the probability-of-collision metric can be misled by this effect, deciding not to maneuver because the probability appears low, even though the elevated uncertainty itself poses the risk. The remedy is to require both that the probability of collision be low and that the position uncertainties be sufficiently small before deciding not to maneuver. Modern conjunction screening tools incorporate this consideration, but the underlying physics ensures that some degree of operator judgment remains necessary.

Commercial Space Situational Awareness

The U.S. Space Surveillance Network has long been the primary source of tracking data and conjunction warnings worldwide. However, commercial space situational awareness providers have grown rapidly over the past decade. Several commercial operators now offer tracking and screening services that complement, and in some cases compete with, the U.S. military capability. The growth of commercial SSA reflects both the limitations of military capabilities in certain operational contexts and the rising commercial demand for higher-precision tracking data.

Founded in 2016 and based in Menlo Park, California, LeoLabs operates a global network of phased-array radars purpose-built for low Earth orbit tracking. The company's radars in Midland, Texas; Fairbanks, Alaska; Costa Rica; Western Australia; New Zealand; and the Azores provide global coverage with update rates markedly higher than those of the U.S. military catalog. The company sells tracking data and conjunction services to satellite operators, regulatory agencies, and other customers. LeoLabs has emerged as a particularly important commercial provider for constellation operators, who require high-update-rate tracking for the dense conjunction environments their constellations create.

NorthStar Earth and Space, based in Montreal, Canada, has developed a different approach using space-based optical sensors deployed in a constellation. The company's first three operational satellites entered service in 2025, providing observations that complement ground-based sensors, particularly at higher altitudes where ground-based optical observation is constrained. The space-based approach offers substantial coverage advantages but requires maintaining the operational satellites, with associated cost and reliability considerations.

ExoAnalytic Solutions, based in Mission Viejo, California, operates a global network of optical telescopes that primarily track the geostationary belt and other high-altitude regions. The company's network of more than 350 telescopes across multiple sites is the most extensive commercial optical tracking capability in operation. Both commercial customers and government agencies widely use the network for high-altitude SSA.

Slingshot Aerospace, based in Austin, Texas, provides analytical and software services that integrate tracking data from multiple sources into operational tools for satellite operators. The company's products include conjunction screening, mission planning, and operational decision support. The company does not operate sensors but has become an integrator of data from other sources, including both government and commercial providers.

Other commercial providers include Aerospace Corporation (a federally funded research and development center that performs SSA functions for the U.S. government and other customers), Comspoc (formerly AGI's Center for Space Standards & Innovation), Space Domain Awareness (a UK-based provider), and smaller companies that specialize in specific capabilities or customer segments. The commercial SSA market has grown exponentially over the past five years. It is expected to continue growing as the satellite population increases and commercial operators seek more capable tracking services.

The relationship between commercial and government SSA providers has evolved over the same period. The U.S. government's transition of conjunction services from military (Space Command) to civilian (NOAA Office of Space Commerce) provision, beginning in 2021, has been accompanied by increased purchases of commercial tracking data to augment government-operated sensors. The European Space Agency has similarly expanded its use of commercial data sources. The trend is toward a hybrid model in which government-operated sensors provide a baseline capability and commercial providers contribute specialized capabilities and additional capacity.

The Eighteenth Space Defense Squadron and Space-Track.org

The Eighteenth Space Defense Squadron, based at Vandenberg Space Force Base in California, manages the day-to-day operations of the U.S. Space Surveillance Network and maintains the principal U.S. catalog of orbital objects. The squadron operates the public-facing space-track.org website, where registered users can access catalog data, conjunction warnings, and related information. The site is the primary means by which the U.S. military's tracking infrastructure serves the broader international community.

Space-track.org provides several categories of data with varying access conditions. The basic catalog of cataloged objects, in TLE format, is available to registered users with relatively few restrictions. Higher-precision orbital data, conjunction warnings, and specific analytical products are available under more restrictive arrangements that reflect both the data's military origin and the U.S. government's expanded commercial engagement. Access arrangements have evolved, generally toward broader availability, but with ongoing restrictions on the most sensitive data.

The Combined Space Operations Center, also based at Vandenberg, coordinates with international partners on joint operations and data sharing. The center's partners include the equivalent operations centers of the United Kingdom, Canada, Australia, France, Germany, New

Zealand, and Norway, with cooperation extending to additional partners through bilateral arrangements. The combined operations approach reflects the recognition that effective SSA requires both technical capability and international coordination, and that no single national capability can fully address the global nature of the orbital environment.

The transition of conjunction services from the U.S. military to NOAA's Office of Space Commerce, mentioned earlier in this chapter, will affect the Eighteenth Space Defense Squadron's role over the coming years. The squadron will continue to operate the Space Surveillance Network and maintain the catalog, but the public-facing services for commercial operators will gradually shift to NOAA. The transition is intended to provide a more commercially friendly interface for civilian users while maintaining the security-relevant aspects of the tracking infrastructure during military operations. The success of the transition depends on NOAA's ability to develop the institutional capability needed to deliver the services the military has historically provided.

Space Traffic Management as a Concept

The term "space traffic management" (STM) has gained use over the past decade to describe a comprehensive approach to the safe and sustainable use of orbital space. The concept extends beyond the technical capabilities of tracking and conjunction screening to include the institutional, regulatory, and operational arrangements that coordinate orbital activity. The analogy is to air traffic management, which combines technical infrastructure (radars, communications, automation) with institutional arrangements (air traffic control centers, operational procedures, regulatory oversight) to coordinate the increasingly complex movement of aircraft through congested airspace.

The space traffic management concept has been the subject of extensive policy discussion in recent years. The U.S. Space Policy Directive 3 (2018) directed the establishment of a comprehensive STM framework under civilian leadership and assigned specific responsibilities to various federal agencies. The European Union has explored its own STM concept through the EU Space Program and related initiatives. The COPUOS LTS Guidelines incorporate STM considerations within their broader framework. The Artemis Accords include provisions for deconflicting activities that address STM concerns.

The substance of space traffic management remains contested in important respects. Some advocates of the concept argue for an active management approach in which authoritative entities issue routing instructions analogous to air traffic control directives. Others argue for a more decentralized approach in which the management system's role is to provide information and coordination services that support operators' autonomous decisions. The latter approach is generally favored by the major commercial operators, who have autonomy in current practice and would resist any framework that constrained their operational decisions through external authority.

The institutional arrangements for space traffic management remain similarly contested. The U.S. has designated NOAA as the lead civilian agency, but the precise allocation of responsibilities among NOAA, DOD, FAA, FCC, and other agencies is still being developed. Internationally, the question of which body should perform STM functions, whether through expansion of existing institutions like ITU or COPUOS, creation of a new dedicated body, or coordination among national authorities, has not been resolved. Institutional development is likely to proceed incrementally, with specific functions assigned as they become operationally necessary, rather than through a comprehensive design of a new institutional architecture.

The technical content of space traffic management includes several well-developed elements and others that remain in their early stages. Conjunction screening, as examined above, is well developed and operational. Coordination among constellation operators on the same orbital shells, addressed through bilateral arrangements among leading operators, is operational but lacks formal institutional support. Long-term traffic forecasting, which would identify orbital regions approaching capacity limits, is in its early stages. Coordination of launch trajectories with on-orbit operations, which would prevent conflicts between ascending vehicles and orbiting satellites, is addressed informally by several launch operators but lacks a comprehensive framework. Each of these elements is being developed by various combinations of government agencies, commercial operators, and academic institutions, and the maturation of the overall STM framework will depend on integrating these development efforts.

Constellation Coordination and Conjunction Avoidance

The deployment of large satellite constellations has significantly expanded the scope and complexity of conjunction screening and avoidance. Each Starlink satellite operates within a constellation of approximately 10,400 to 10,500 active satellites, with elevated conjunction risk both within the constellation and with other operators' assets. SpaceX has reported that Starlink satellites performed tens of thousands of avoidance maneuvers in 2023 and 2024, with the rate climbing as both the constellation and the surrounding debris environment have grown. The OneWeb constellation, with approximately 630-650 satellites in higher orbits, faces a different but substantial conjunction environment. Amazon Leo, with deployment progressing through 2026, will add another large constellation operating at altitudes that overlap with portions of the existing population.

SpaceX's publicly released figures put the 2023 total at approximately 25,000 maneuvers across the Starlink constellation, rising to approximately 50,000 in 2024 as both the constellation and the surrounding debris environment grew. Translated into a per-satellite figure, this amounts to roughly four to six avoidance maneuvers per satellite per year, with the rate varying significantly by orbital shell and proximity to dense conjunction zones. These figures represent a genuine operational burden. Each maneuver consumes between 0.1 and 0.5 meters per second of delta-V from the satellite's propellant budget, cumulatively reducing the propellant available for end-of-life disposal. SpaceX has reported that its disposal planning accounts for this drawdown, but the margin is not unlimited and narrows as the conjunction rate rises.

The major constellation operators have invested heavily in autonomous conjunction-avoidance capabilities. Starlink satellites carry onboard software that evaluates conjunction warnings and, in routine cases, executes avoidance maneuvers without ground intervention. Autonomy reduces the operational burden on ground controllers and improves response speed, particularly for warnings issued shortly before closest approach. This autonomy is bounded by configuration parameters set by the operator, including the probability thresholds at which automatic maneuvers are executed and the categories of conjunctions that require human review. Other constellation operators have developed similar capabilities to varying degrees, reflecting an industry trend toward greater autonomy.

Coordination among constellation operators is increasingly important as the orbital population grows. Bilateral arrangements among leading operators, such as Starlink's with OneWeb and Amazon Leo's with Starlink, establish protocols for sharing operational information and coordinating maneuvers. These arrangements typically specify which operator has priority in specific conjunction scenarios, how operational data is shared, and how disputes are resolved. The arrangements are voluntary and depend on operators' willingness to cooperate; large commercial operators have generally found cooperation to be in their mutual interest, but smaller operators or those based in jurisdictions with limited regulatory oversight may not participate in similar arrangements.

The Space Safety Coalition, discussed in Chapter 8, has developed best practices for constellation operations, including specific guidance on conjunction avoidance and coordination. The coalition's membership includes major commercial operators across multiple jurisdictions, and its guidance has been adopted by many operators as the basis for their operational practices. The coalition's framework is voluntary, limiting its direct enforcement authority but providing a useful normative framework that has shaped industry practice.

The Chinese constellation programs, including the Guowang and Qianfan constellations, operate under different institutional arrangements that have not been the subject of detailed international examination. Whether Chinese operators participate in the bilateral coordination arrangements that have developed among Western commercial operators, and whether their conjunction-avoidance practices align with emerging international norms, remain open questions whose answers will affect the operational debris environment as the Chinese constellations grow.

The Limits of Current Capability

The current global tracking infrastructure has serious limitations that constrain the management of the orbital debris environment. Several specific limitations warrant discussion.

The cataloging threshold remains around 10 centimeters (3.9 inches) in low Earth orbit, despite advances in capability in recent years. Objects smaller than this threshold contribute meaningfully to collision risk; an object a few centimeters (inches) in size can disable a functional satellite but cannot be tracked individually. As a result, conjunction screening can address only collisions with cataloged objects. In contrast, the risk posed by smaller objects must be addressed statistically through shielding design and other approaches that do not rely on real-time tracking.

The position uncertainties for cataloged objects remain large enough to cause substantial operational complications. The dilution problem discussed above is one consequence. More broadly, the conjunction screening process generates many warnings with high uncertainty, requiring operator judgment on whether to maneuver. Improving position accuracy depends on deploying more capable sensors and developing better analytical techniques, both of which are progressing. Still, neither is likely to eliminate the underlying uncertainty in the near term.

The catalog completeness for high-altitude regions, particularly the medium Earth orbit and geostationary regimes, remains less comprehensive than for low Earth orbit. Atmospheric and geometric factors limit optical observations of these regions, and space-based sensors to address them are still in the early stages of deployment. As a result, the conjunction screening process is most effective in low Earth orbit and progressively less effective at higher altitudes.

A related but distinct limitation is catalog saturation during fragmentation events. When a large object breaks apart, it can produce thousands of fragments within minutes. The tracking network must observe, process, and assign catalog numbers to each fragment while maintaining custody of the existing 47,000 cataloged objects. The 2007 Fengyun-1C event produced fragments so quickly that it took the U.S. Space Surveillance Network more than three weeks to catalog most trackable fragments. During that window, operators could not reliably screen their satellites against the new population. The 2021 Cosmos 1408 event produced a similar temporary degradation in screening quality. As the overall population grows, each new fragmentation event will be harder to catalog quickly, and the window of uncertainty after a large fragmentation event will likely lengthen rather than shorten unless tracking infrastructure scales faster than the threat itself.

The institutional arrangements for international coordination are less developed than the technical capabilities. The bilateral agreements between the U.S. and other states, the various commercial arrangements, and the multilateral discussions in COPUOS provide the principal current mechanisms. Still, no single institution has clear authority for international coordination. As a result, operational coordination depends on the voluntary cooperation of multiple actors and has limited capacity to resolve disputes or enforce common practices.

The legal framework governing tracking and SSA activities remains unclear in several critical areas. The position of objects in space has not been treated as private information in legal frameworks developed for other domains. However, operators' operational concerns include both the security of their satellites and the commercial sensitivity of their practices. Various jurisdictions have begun to address these issues through their domestic legal frameworks, but no comprehensive international framework exists.

The Emerging Global Picture

Despite the limitations identified above, the global tracking infrastructure has advanced rapidly over the past decade and continues to mature. Several developments will shape its trajectory in the coming years.

The Space Fence at Kwajalein, which entered operational service in 2020, has expanded U.S. capability to track small objects in low Earth orbit. The system can detect objects as small as about 5 centimeters (2 inches) at altitudes up to 1,500 kilometers (932 miles), thereby expanding the catalog of small objects. The system's contribution to understanding the debris environment has been substantial. However, specific findings on the population of small objects have been published incrementally rather than in a single comprehensive release.

ESA's Space Safety Program has expanded European tracking capabilities by deploying new sensors at multiple sites and acquiring commercial data. The program has supported the development of conjunction screening services for ESA missions and European national programs, providing an alternative to U.S. military services for European users.

The commercial SSA sector continues to grow, with multiple new entrants and significant investments in capabilities in recent years. Growth has been supported by both commercial demand from satellite operators and government purchases of commercial data to augment government-operated sensors. The trend toward a hybrid model integrating government and commercial capabilities is likely to continue.

International cooperation on tracking has expanded through bilateral agreements between the U.S. and other states, coordination by the Combined Space Operations Center with allied operations centers, and broader engagement by various states with the international SSA community. The cooperation is uneven; Russia and China participate at much lower levels than allied states, but the overall trend has been toward greater international engagement.

The institutional arrangements for SSA continue to evolve. The U.S. transition from military to civilian provision of conjunction services, the European development of operational SSA services through ESA, and the development of national capabilities in the United Kingdom, Australia, Japan, and other states represent a broader pattern of institutional maturation that has not yet stabilized but points toward a more comprehensive global infrastructure than has existed at any prior point.

Future Trajectories for Tracking and Space Traffic Management

Several factors will shape the trajectory of the global tracking infrastructure over the coming decade. The growth of the satellite population, particularly through continued constellation deployment, will require a corresponding increase in tracking capacity. The development of new sensors, including next-generation phased-array radars, space-based optical sensors, and advanced analytical capabilities, will expand both the volume and the precision of tracking data. The institutional arrangements for international coordination will need to evolve to match technical capabilities, with particular attention to integrating commercial capabilities and coordinating activities across multiple jurisdictions.

Several specific developments are anticipated. The deployment of additional space-based optical sensors by NorthStar, the U.S. Space

Force, and other operators will expand coverage of high-altitude regions. The maturation of artificial intelligence applications for conjunction screening, addressed in more detail in Chapter 12, will improve the speed and accuracy of conjunction analysis. The expansion of commercial conjunction services to a wider range of operators will enhance the operational capability of smaller commercial actors. The development of more sophisticated automated avoidance capabilities will reduce the operational burden on ground controllers.

The challenges facing infrastructure are substantial, however. The growth of the cataloged population will continue to outpace tracking capacity for the foreseeable future, requiring ongoing efficiency improvements to maintain the current level of service. The increasing share of small objects, which contribute heavily to collision risk but cannot be individually tracked, will require continued development of statistical approaches to manage the risk they pose. Operating across multiple jurisdictions with diverse regulatory frameworks and operational practices will require ongoing institutional development to ensure effective coordination.

The deeper question for the long-term trajectory is whether the tracking and SSA infrastructure can scale fast enough to support the growth of orbital activity. The current infrastructure was designed for an environment with a few thousand active satellites and a manageable conjunction load. The contemporary environment has approximately 11,000 active satellites and a conjunction load that has grown dramatically. The projected environment for the next decade includes potentially 50,000 or more active satellites and a conjunction load that would far exceed the current infrastructure's capacity. Whether the infrastructure can scale to meet this demand and whether the institutional arrangements can evolve quickly enough to support that scaling are open questions that will strongly shape the trajectory of the orbital debris environment.

Figure 9.1. Global Distribution of Selected U.S. Space Surveillance Network Sensors, 2026

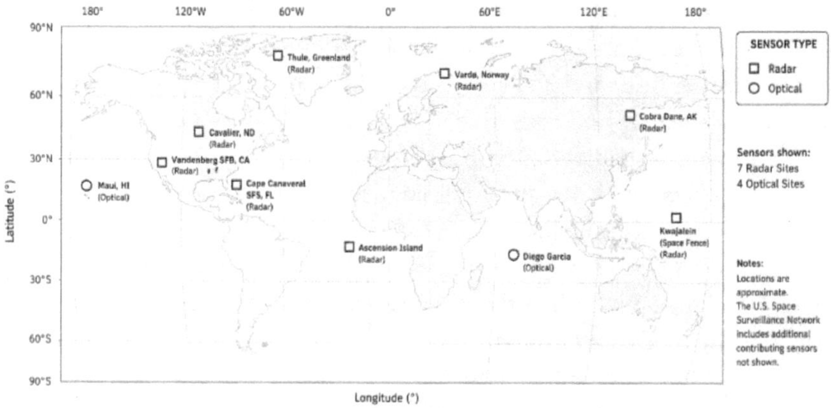

RADAR SITES	LATITUDE	LONGITUDE		LATITUDE	LONGITUDE	OPTICAL SITES	LATITUDE	LONGITUDE
Cobra Dane, AK	55.1° N	162.7° W	Vandenberg SFB, CA	34.7° N	120.6° W	Maui, HI	20.8° N	156.3° W
Thule, Greenland	76.5° N	68.7° W	Cape Canaveral SFS, FL	28.5° N	80.6° W	Diego Garcia	7.3° S	72.4° E
Vardø, Norway	70.4° N	31.1° E	Ascension Island	7.9° S	14.4° W			
Cavalier, ND	48.8° N	97.4° W	Kwajalein (Space Fence)	8.7° N	167.7° E			

Figure 9.2. Conjunction Probability vs. Miss Distance: The Dilution Problem in High-Uncertainty Regimes

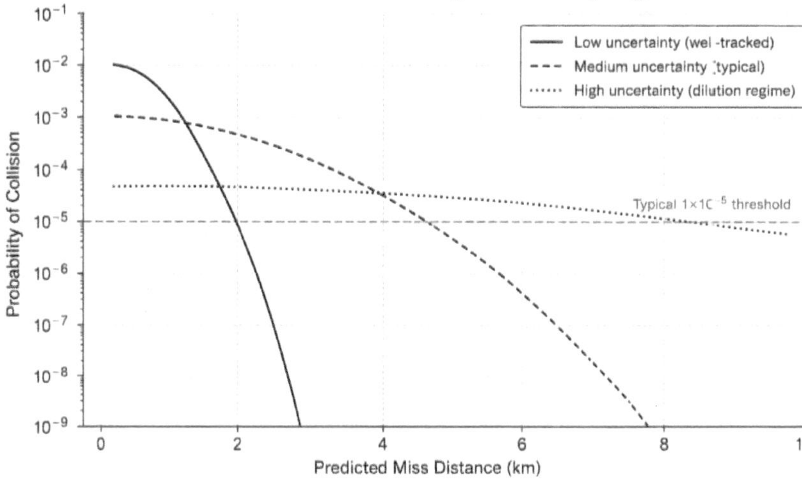

Sensor	Type	Location	Notes
AN/FPS-85	Phased-array radar	Eglin AFB, FL	Primary LEO tracker
Cobra Dane	Phased-array radar	Shemya, AK	Polar coverage
Globus II	Mechanical radar	Vardø, Norway	High-latitude LEO
Space Fence	S-band phased-array	Kwajalein Atoll	Operational 2020; ~5 cm at 1,500 km ~2 in. at 932 miles
GEODSS	Optical	Socorro NM, Maui HI, Diego Garcia	High-altitude tracking
Space-Based Surveillance System	Optical	Low Earth orbit	High-altitude observation

Table 9.1. Principal Sensors of the U.S. Space Surveillance Network

Provider	HQ	Approach	Notable Customers
LeoLabs	Menlo Park, CA	Global radar network	Constellation operators, regulators
NorthStar Earth and Space	Montreal, Canada	Space-based optical	Government, commercial
ExoAnalytic Solutions	Mission Viejo, CA	Optical telescope network	GEO operators, U.S. government
Slingshot Aerospace	Austin, TX	Software & data integration	Operators, regulators
ComSpOC (Aerospace Corp)	Various	Federally funded R&D center	U.S. government, allies

Table 9.2. Commercial Space Situational Awareness Providers, 2026

Chapter Summary

The global infrastructure for tracking, space situational awareness, and space traffic management is the empirical foundation on which all aspects of debris policy rest. Over six decades, it has evolved from a handful of military radars into a globally distributed network of sensors, data systems, and analytical capabilities that process hundreds of thousands of observations per day. The U.S. Space Surveillance Network maintains the principal global catalog of orbital objects, with extensive complementary capabilities operated by Russia, China, ESA, and commercial providers such as LeoLabs, NorthStar, and others. Conjunction screening, performed primarily through the U.S. Combined Space Operations Center and commercial providers, generates warnings of close approaches that operators use to decide on avoidance maneuvers. The space traffic management concept, while still evolving in its institutional and operational specifics, provides the framework within which these activities are increasingly coordinated. The current infrastructure has substantial limitations, including cataloging thresholds, positional uncertainties, institutional fragmentation, and incomplete international coordination, but it has advanced over the past decade and continues to mature. The challenge facing the infrastructure is whether it can scale to meet the growth of orbital activity while developing the institutional arrangements that effective coordination requires. The chapters that follow examine the operational, technical, and economic systems that depend on this infrastructure: mitigation and prevention (Chapter 10), active debris removal (Chapter 11), and the broader integration of debris considerations into the economics and sustainability of space activity (Chapters 13 and beyond).

Chapter 10. Mitigation: Prevention and Design Standards

Introduction

Mitigation, the prevention of new debris generation through the design and operation of spacecraft, is the foundation of any plausible response to the orbital debris problem. The cascade analysis in Chapter 4 identifies the conditions under which the existing population could grow through internal collisions. However, stability requires that the rate of new debris generation be controlled. Active removal, examined in Chapter 11, addresses the existing population but cannot solve the problem if new debris is generated faster than it can be cleared. The discipline of mitigation, applied consistently across the international space community, is therefore necessary, though not sufficient, for any successful response to the debris challenge.

This chapter examines the technical and operational practices that constitute mitigation: end-of-life disposal of satellites and rocket stages; passivation to prevent post-mission explosions; preventing the release of operational debris during normal operations; conjunction avoidance to prevent accidental collisions; and design measures to reduce the consequences of failures that occur despite preventive measures. The chapter also examines the international and national standards that codify these practices, the compliance challenges operators face in implementing them, and the evolving frontiers of mitigation practice as the orbital environment changes. The treatment is sufficiently detailed to support the analysis in later chapters of the specific technologies and economic frameworks within which mitigation operates.

End-of-Life Disposal

The disposal of satellites and rocket stages at the end of their operational lives is the most consequential mitigation practice. A satellite that remains in orbit indefinitely after mission completion contributes to the debris population both directly, as a large object that occupies orbital space and can be struck by other objects, and indirectly, as a potential source of thousands of fragments if it is struck. The systematic disposal of satellites within defined timeframes after mission completion is therefore the principal mechanism by which the operational satellite population is prevented from growing into a long-term debris population.

The disposal options vary by altitude. For satellites in low Earth orbit, the principal option is reentry into the atmosphere, either through controlled deorbit using the satellite's propulsion system or through atmospheric drag acting on satellites that have completed their mission and been left in orbit. For satellites in higher orbits, where atmospheric drag is too weak to remove them within reasonable timeframes, the principal option is transfer to a graveyard orbit, an orbital regime designated for disposal where the satellite poses no risk to operational missions. For satellites at intermediate altitudes, both options may be technically feasible, with the choice depending on operational considerations, including propellant availability and operator preference.

The disposal timeline has been one of the most actively debated aspects of mitigation policy. The IADC mitigation guidelines of 2002 established a twenty-five-year disposal timeline as the international baseline, based on technical analysis of how rapidly debris populations grow under different disposal strategies. The 25-year baseline was retained in the 2021 revision of the guidelines but has been progressively superseded by tighter national standards. The U.S. FCC's 2022 five-year rule, the UK's 2021 five-year rule, and the ESA's five-year requirement for most missions have established a new, effective baseline among the major spacefaring states. Whether the international guidelines will be revised to reflect this new national baseline remains an open question.

The technical means of disposal depend on the satellite's design and operational capability. Satellites equipped with propulsion systems can perform controlled deorbit burns that place them on a trajectory leading to atmospheric reentry within a defined period. The propellant required for a controlled deorbit from low Earth orbit is typically 5 to 10 percent of the satellite's launch mass, a substantial allocation that affects both cost and satellite design. Satellites without propulsion systems, or those that fail to complete their disposal maneuvers, rely on atmospheric drag for eventual reentry. The natural decay timeline depends on the satellite's altitude, mass, and cross-sectional area; for satellites at altitudes below approximately 600 kilometers (373 miles), decay typically occurs within decades, while satellites at higher altitudes may persist for centuries.

The compliance challenges in end-of-life disposal are significant. Disposal failures can arise from multiple mechanisms: propulsion failures that prevent the deorbit maneuver; attitude control failures that prevent precise targeting of the disposal trajectory; software failures that disrupt the disposal sequence; communication failures that prevent the operator from commanding disposal; and structural failures that disable the satellite before disposal can be initiated. The historical record includes many examples of each failure mode. The cumulative effect is that compliance with disposal requirements consistently falls below 100 percent, with the precise rate varying by operator, satellite type, and operational environment.

For low Earth orbit constellations, the disposal challenge is intensified by the scale of operations. A constellation of 10,000 satellites with a five-year average lifetime implies that approximately 2,000 satellites must be disposed of each year through reliable, automated processes. The reliability of those processes determines the constellation's long-term contribution to debris. A 95 percent compliance rate, typically considered good performance for individual missions, would leave 100 satellites per year in orbit beyond their disposal timelines, for a total of 1,000 satellites over a decade. Modeling studies have examined the implications for the long-term debris environment, consistently concluding that it is sensitive to disposal performance at the percent level. Constellation operators have therefore invested in disposal reliability, with the major commercial constellations reportedly achieving compliance rates that exceed 99 percent in normal operations.

Disposal Failures and Their Consequences

The historical record includes detailed information on disposal failures and the policy responses they have prompted. Several specific cases illustrate these patterns and the lessons drawn from them.

The Envisat satellite, launched by ESA in 2002 and operational until communications were lost in April 2012, is the largest single satellite in low Earth orbit and a significant debris hazard. The satellite has a mass of approximately 8 metric tonnes (8.8 tons) and orbits at an altitude of approximately 770 kilometers (478 miles), high enough that natural decay would not occur for centuries. ESA had not designed the satellite for active deorbit; the loss of communications eliminated the possibility of any later remediation. Envisat is now a defunct, uncontrolled object with a high risk of collision in one of the most heavily affected altitude bands. The satellite has been identified as a high-priority target for active debris removal in multiple analyses, and ESA's ClearSpace-1 mission was originally designed to target Envisat before being retargeted to PROBA-1 in 2024.

The Envisat case illustrates several lessons. First, satellites operating at altitudes where natural decay does not occur must be designed for active disposal; the absence of disposal capability creates long-term debris liabilities that ordinary mitigation cannot address. Second, the design choices made during mission planning have consequences that extend for decades after the mission ends; the failure to require disposal capability in 1990s-era satellites has created problems that the current generation of operators must manage. Third, the cost of remediation, through active debris removal or other means, is far greater than the cost of designing for proper disposal in the first place. These lessons have been incorporated into modern design practice for satellites operating at high altitudes, but the legacy population from earlier eras remains a serious concern.

The 2015 fragmentation of NOAA-16 illustrates a distinct category of disposal failure. The satellite had operated successfully through 2014, when it was decommissioned, with its batteries and propellant tanks left in their operational configuration rather than passivated. A subsequent battery failure in November 2015 caused the satellite to fragment, producing approximately 275 cataloged fragments at altitudes near 850 kilometers (528 miles). The fragmentation could have been prevented through proper passivation at decommissioning; the failure to passivate was a procedural failure rather than a design limitation. The episode prompted NOAA and other operators to review their decommissioning procedures and implement more systematic passivation requirements.

The legacy population of unpassivated upper stages, particularly Russian and earlier Soviet stages, continues to produce fragmentation events at intervals of years to decades. Each event produces hundreds to thousands of fragments. The cumulative contribution of these events over the past several decades has been significant, comparable to the contribution of anti-satellite tests. Depleting the legacy population, either through the eventual explosion of all unpassivated stages or proactive removal, will be a considerable step toward stabilizing the long-term debris environment. Still, the timeline extends for decades into the future.

Passivation

Passivation is the systematic removal of stored energy from spacecraft and rocket stages at the end of the mission, eliminating the primary sources of post-mission explosions. The primary forms of stored energy in spacecraft and stages are propellant in pressurized tanks, electrical energy in batteries, and mechanical energy in pressurized gas systems. Each can cause explosive failure if containment fails, and the historical record includes many examples of fragmentation events caused by each.

Propellant passivation is performed by venting residual propellant after the operational mission is complete. For monopropellant systems, the propellant can be expelled through the spacecraft's existing propulsion thrusters during an end-of-life burn. For bipropellant systems, both fuel and oxidizer must be vented, and the venting order and procedures must be designed to prevent inadvertent reactions during venting. For pressurized propellant systems, the gas pressure must also be vented to prevent subsequent pressure-driven failures of the tank structure.

The technical requirements for effective passivation have evolved over the past three decades. The first systematic attention to passivation in the U.S. context arose in response to the Delta upper-stage explosion problem of the 1970s, which led to a series of fragmentation events involving spent stages left in orbit with residual propellant. The response was the development of passivation procedures that became standard practice for U.S. upper stages by the late 1990s. These procedures have been refined since then and are now applied routinely for U.S. and ESA stages.

The Russian and Chinese launch vehicle programs have applied passivation in more varied ways. Many newer Russian and Chinese stages incorporate passivation procedures that meet Western standards, but the legacy population of older stages includes many that lack effective passivation. The ongoing pattern of explosions from older Russian stages, including Briz-M events in recent years, illustrates the enduring consequences of the historical absence of passivation discipline. Similarly, the 2022 Long March 6A breakup discussed in Chapter 5 shows that passivation discipline is not uniformly applied even in newer Chinese stages.

Battery passivation is a more recent focus in mitigation practice. The NOAA-16 fragmentation discussed above, caused by battery failure in a defunct satellite, was one of the events that prompted greater attention to battery passivation. The basic technique is to discharge batteries to safe states before mission end, eliminating the chemical energy that could drive explosive failures during later thermal cycling. The technique requires specific procedures tailored to each battery chemistry and has been progressively incorporated into mission planning for modern satellites. Older satellites that did not implement battery passivation continue to pose a small but sustained risk of fragmentation events from this category.

Other categories of stored energy include pressurized gas in attitude control systems, kinetic energy in rotating mechanisms (such as reaction wheels and gyroscopes), and chemical energy in pyrotechnic devices. Each category has been progressively addressed through targeted design and operational practices. The cumulative effect of comprehensive passivation discipline is to reduce the rate of post-mission fragmentation events significantly. However, the legacy population from earlier eras continues to generate events at lower rates than would occur without passivation.

Operational Debris Release

A category of debris generation distinct from explosions and collisions is the deliberate or inadvertent release of mission-related items during normal operations. Historical examples include the deliberate jettisoning of equipment during space station operations, the release of cover materials and protective devices during satellite deployment, the loss of equipment by astronauts during extravehicular activities, and the inadvertent release of small components due to structural failures or operational accidents. Each of these categories has contributed to the historical debris population and has been progressively addressed through changes in operational practice.

The deliberate jettisoning of equipment from space stations was a common historical practice that has been reduced in modern operations. The Mir space station, which operated from 1986 to 2001, jettisoned numerous items during routine maintenance. The International Space Station has adopted more disciplined practices, returning most discarded items to Earth via cargo vehicles rather than jettisoning them into orbit. Notable exceptions have occurred; however, the 2024 ISS battery pallet reentry, which produced fragments recovered in Florida, resulted from a 2021 jettison operation, which had been justified as the only practical means of disposing of the bulky equipment.

The release of deployment-related debris has steadily decreased due to design improvements in modern satellites. Spring-actuated separation mechanisms have replaced explosive bolts in many applications, eliminating the small pyrotechnic debris the bolts produced. Cover materials and protective devices that were historically jettisoned at deployment are now retained on the spacecraft or designed for in-flight stowage. The cumulative effect is that modern satellite deployments produce far less mission-related debris than was typical in earlier eras.

The loss of equipment during extravehicular activities has been a minor, ongoing source of debris in space station operations. Items lost during EVAs, including tools, components, and occasional equipment, generally enter orbits similar to the station's and decay through natural drag within months to years. The cumulative debris contribution is small, but operational practices have been progressively refined to reduce the rate of losses. Tethering protocols, equipment design improvements, and operational training have all contributed to this reduction.

The inadvertent release of components due to structural failures or operational accidents is more difficult to address through design or operational practices. Each event is, by its nature, unanticipated, and the lessons drawn from individual events apply primarily to preventing similar future events rather than to a categorical solution. The cumulative contribution of these events is small but steady, and modeling studies suggest it will remain at low rates for the foreseeable future.

Conjunction Avoidance

Active conjunction avoidance, the maneuvering of operational satellites to avoid predicted close approaches with other objects, has become an essential operational practice for satellites in densely populated orbital regions. This practice depends on the conjunction screening infrastructure discussed in Chapter 9 and on the propulsion and operational capabilities of individual satellites. Across the satellite population, the cumulative effect is the prevention of additional Iridium-Cosmos-style accidents that would substantially increase the debris population.

The technical practice of conjunction avoidance involves several distinct elements. Conjunction warnings are received from the screening infrastructure, either the U.S. Combined Space Operations Center or commercial providers, and indicate predicted close approaches. These warnings are evaluated based on the predicted miss distance, position uncertainties, the probability of collision, and the operator's risk tolerance. If the conjunction warrants a maneuver, the maneuver is planned to achieve sufficient separation while minimizing operational impact. The maneuver is executed by the satellite's propulsion system, with verification through subsequent tracking.

The frequency of conjunction avoidance maneuvers varies widely by satellite. The International Space Station has performed dozens of avoidance maneuvers over its operational lifetime, with the rate increasing as the surrounding debris environment has grown. Commercial constellation satellites perform maneuvers more frequently because of the dense conjunction environment within the constellations and the volume of cataloged objects in their orbital regions. SpaceX has reported tens of thousands of avoidance maneuvers across the Starlink constellation in recent years; the rate averages several maneuvers per satellite per year, with wide variation across the constellation depending on local conjunction environments.

The cumulative cost of conjunction avoidance maneuvers is burdensome for major operators. Each maneuver consumes propellant that would otherwise extend operational life. Each maneuver may briefly interrupt operations, with corresponding impacts on service. Across constellations of thousands of satellites, the aggregated cost is high, though it remains small relative to the value of the services provided. As the debris environment grows, the cost rises, and the question of when it becomes prohibitive, when the operational impact of conjunction avoidance approaches the value of operating in the affected orbital environment, is one of the questions that the cascade analysis frames.

The autonomous conjunction-avoidance capabilities developed by leading constellation operators mark an important technical advance. The Starlink constellation's onboard autonomous response to conjunction warnings reduces the operational burden on ground controllers and improves response speed, particularly for warnings issued shortly before closest approach. This autonomy is bounded by configuration parameters set by the operator, including the probability thresholds at which automatic maneuvers are executed and the categories of conjunctions that require human review. Across the industry, the trend is toward greater autonomy, with the operational practices of the largest constellation operators serving as the model for smaller operators.

Coordinating conjunction avoidance among operators is increasingly important as the satellite population grows. When two active satellites are predicted to encounter each other, the question of which satellite should maneuver and which should remain on its trajectory must be resolved. The historical practice, based on operator preference and bilateral negotiation, has been progressively replaced by formal coordination protocols that establish priorities based on satellite characteristics, mission status, and other factors. The Space Safety Coalition's best practices include guidance on conjunction coordination, and the major operators have developed bilateral protocols that set specific procedures. The voluntary nature of these arrangements limits their enforcement capability but provides a useful operational framework.

Design for Reduced Debris Impact

Beyond operational practices for disposal, passivation, and conjunction avoidance, modern spacecraft design incorporates specific measures intended to reduce the debris consequences of failures that occur despite preventive measures. These measures fall into several categories.

Design for demise, discussed in Chapter 6, involves deliberately selecting materials and configurations to maximize the fraction of mass that disintegrates during reentry. This practice reduces the risk of casualties from uncontrolled or partially controlled reentries and complements the controlled reentry capability of some satellites. It has been adopted by some satellite manufacturers, particularly for satellites operating at altitudes where natural decay clears them within applicable timelines. Implementation is most straightforward for small satellites, where structural requirements permit significant flexibility in material choice; larger satellites typically have less flexibility, making design for demise correspondingly more difficult.

Fragmentation reduction refers to design practices that minimize the number of fragments produced when a satellite breaks up, whether through accidental fragmentation or controlled disposal. These practices include avoiding large external structures that could generate many fragments if struck, using designs that fail into fewer, larger fragments rather than more, smaller ones, and selecting materials that produce predictable fragmentation patterns. These practices remain in the early stages of standardization but have been incorporated into specific mission designs.

Damage tolerance is a satellite's ability to survive small-fragment impacts without a complete loss of mission. This practice is most relevant to large satellites, where a complete loss would have serious consequences. Damage-tolerant designs incorporate redundancy, structural margins, and operational practices that allow the satellite to continue functioning after sustaining damage. This approach has been developed for human spaceflight applications, where the safety implications of damage are most severe. For uncrewed satellites, the cost-benefit analysis of damage tolerance must be weighed against alternative risk-reduction approaches.

Shielding involves using structural elements designed to protect critical components from small-fragment impacts. The most mature shielding designs are used in crewed spacecraft, particularly the International Space Station's multi-layer Whipple shields and stuffed Whipple shields. These designs are based on hypervelocity impact test data and protect against fragments up to defined size thresholds at specified impact velocities. Smaller satellites typically cannot afford the mass penalty of comprehensive shielding and instead rely on reduced exposure (smaller cross-sections, shorter operational lifetimes) for protection.

Standards and Guidelines

The mitigation practices discussed above are codified in a hierarchy of standards and guidelines, ranging from international consensus documents to specific national requirements. The relationships among these documents have been examined in Chapters 7 (international level) and 8 (national level); this section examines the technical content as it specifically addresses mitigation.

The IADC Space Debris Mitigation Guidelines, in their 2002 original and 2021 revision, establish the technical baseline for mitigation. The guidelines address the primary categories of debris generation and provide specific recommendations for each. The 2021 revision addressed several issues that had emerged since the 2002 original, including guidance for large constellations, a more detailed treatment of post-mission disposal, and clarification of the relationship between mitigation and active debris removal.

ISO 24113, the standard for space systems debris mitigation, provides a more detailed technical framework that operationalizes the IADC guidelines. It includes specific compliance verification procedures, more detailed technical specifications, and guidance on applying the principles to specific mission types. The standard has been adopted as the basis for compliance verification across multiple regulatory frameworks and serves as a common technical reference across jurisdictions.

The COPUOS Long-term Sustainability Guidelines of 2019 provide a high-level policy framework that incorporates debris mitigation alongside other sustainability considerations. The guidelines are less technically detailed than the IADC guidelines or ISO 24113, but they provide the political endorsement that supports the broader implementation of the technical standards.

National standards, examined in Chapter 8, vary in their specific requirements but generally derive from the IADC framework. The major differences concern disposal timelines (25 years versus 5 years), casualty risk thresholds (1-in-10,000 versus 1-in-100,000), and design requirements for preventing fragmentation. The trend toward national standards more stringent than the international baseline has been a principal development in mitigation policy over the past five years.

Implementing standards through compliance verification involves several distinct activities. Pre-launch review confirms that the operator has documented the mission's compliance with applicable standards. Post-launch monitoring confirms that operational practices align with the documented commitments. End-of-life verification confirms that disposal commitments have been met. The rigor of each verification step varies widely across jurisdictions, with the U.S. and ESA frameworks generally applying the most rigorous practices.

Compliance Challenges and Solutions

The implementation of mitigation standards faces several persistent compliance challenges that have shaped the development of operational practices.

Disposal failures, discussed earlier in this chapter, are the most consequential compliance issue. Even the best-designed satellites can fail to be disposed of successfully, and the cumulative effect of these failures across thousands of satellites is substantial. The principal solutions involve design redundancy, operational margins, and active monitoring. Redundant propulsion systems enable disposal even if the primary system fails. Operational margins reserve sufficient propellant for disposal, even after extended mission durations. Active monitoring identifies satellites approaching the end of life and ensures that disposal preparations are initiated in time.

Passivation compliance has improved significantly over the past two decades, but it remains uneven. The challenges include legacy populations of older satellites and stages not designed for passivation, operational complications that prevent passivation in certain failure scenarios, and inconsistent application of passivation procedures across operators. The principal solutions involve design improvements (modern stages and satellites are designed with passivation as a baseline), operational discipline (procedures that specifically address passivation requirements), and the progressive replacement of legacy systems.

Conjunction-avoidance compliance depends on the operator's ability to receive and respond to warnings. For small operators with limited operational infrastructure, capabilities may be constrained, and conjunctions that would prompt maneuvers by larger operators may go unaddressed. The principal solutions include commercial conjunction services that provide small operators with capabilities on a fee-for-service basis, regulatory requirements that mandate minimum conjunction-avoidance capabilities, and industry coordination mechanisms that share best practices.

Operational debris release compliance requires both design discipline (preventing the design of systems that generate operational debris) and operational discipline (preventing operational practices that produce debris). The principal solutions include design review processes that catch debris-generating features before they are incorporated, operational training that emphasizes debris awareness, and continuous improvement processes that identify and address new sources of debris as they emerge.

The cumulative effect of these compliance practices, applied across the international space community, has measurably reduced the rate of new debris generation per unit of orbital activity. However, the total rate of debris generation continues to grow because orbital activity has increased faster than per-unit improvements have offset it. The ongoing tension between absolute growth in activity and relative improvement in mitigation discipline remains one of the central challenges of contemporary debris policy.

The numbers behind this improvement are worth stating directly. ESA's annual Space Environment Reports have tracked compliance with the IADC twenty-five-year disposal guideline since the early 2010s. In 2010, roughly 20 to 25 percent of satellites reaching the end of life in low Earth orbit were disposed of in accordance with the guidelines. By 2020, that figure had risen to approximately 50 percent. By 2024, compliance approached 60 percent across the broader satellite population, though the largest commercial constellation operators reported internal compliance rates above 95 percent. The gap between the leading commercial operators and the broader satellite population reflects the difference between operators that have designed, funded, and staffed reliable disposal operations and those that have not. Closing that gap is the central challenge of contemporary mitigation policy.

Mitigation for Large Constellations

The deployment of large constellations has created specific mitigation challenges that the historical mitigation framework was not designed to address. These challenges arise from both the scale of the operations, with thousands of satellites operated by single entities, and the specific operational characteristics of constellation business models.

The disposal challenge for constellations is particularly acute because of the volume of dispositions that constellation operations generate. As discussed earlier, even small failure rates can lead to large numbers of failed satellites over extended operational periods. The major commercial constellations have invested heavily in disposal reliability, but the long-term performance of these investments depends on factors that cannot be fully verified in advance. Whether the constellations will maintain their high reliability over decades of operation, through changes in operating personnel, modifications to satellite designs, and varying economic conditions, remains a question that only experience will answer.

The conjunction environment within constellations is particularly dense, requiring sophisticated coordination of intra-constellation conjunction avoidance. The major operators have developed specific practices for this coordination, including dedicated orbital shells that minimize cross-shell conjunctions, careful phase management to prevent within-shell encounters, and autonomous response capabilities to resolve conjunctions without ground intervention. These practices have been developed primarily for the specific characteristics of each constellation; generalizing them across operators is in its early stages.

Constellations also face new categories of mitigation considerations that smaller missions did not. The astronomical impact of constellations, examined in Chapter 15, is one example: the visibility of large numbers of satellites affects ground-based astronomy in ways the previous mitigation framework did not address. The atmospheric impact of constellation reentries, examined in Chapter 6, is another. Both of these issues are driving the development of new mitigation practices specific to constellation operations.

The regulatory framework for constellations has been evolving to address these challenges. The U.S. FCC's 2022 five-year disposal rule was motivated in part by constellation considerations. The ESA framework has incorporated specific guidance for constellations. The COPUOS LTS Guidelines include provisions related to constellation operations. The development of constellation-specific mitigation frameworks is likely to continue as operational experience with constellations grows.

Frontiers of Mitigation Practice

Several emerging frontiers in mitigation practice warrant particular attention. Each represents an area where current practice is in its early stages and where significant development is expected over the coming years.

Cross-jurisdiction coordination of mitigation practices is one frontier. Although national mitigation standards have narrowed over the past decade, considerable variation remains, creating regulatory complexity for international operators and uneven debris contributions across jurisdictions. The development of mutual recognition arrangements, common verification frameworks, and coordinated tightening of standards is under active discussion. Progress has been slow but steady; whether it will accelerate to match the growth of orbital activity is an open question.

Mitigation for in-orbit servicing and active debris removal is another frontier. The operations of servicing and removal vehicles raise debris-related considerations that the conventional mitigation framework was not designed to address. Specific provisions for the safety of rendezvous and proximity operations, the handling of potentially fragmentation-prone targets, and the disposal of removal vehicles after their missions are under development. The CONFERS organization and several national regulatory bodies have begun developing specific provisions for these operations.

Mitigation for cislunar operations, examined in more detail in Chapter 16, represents an important new frontier. The conventional mitigation framework was designed for Earth-orbiting operations, and its provisions translate imperfectly to the cislunar regime. Developing specific provisions for cislunar mitigation will be one of the principal challenges of debris policy over the coming decade.

Real-time mitigation, in which mitigation decisions are operationally made based on current conditions rather than on pre-launch design, is a developing concept. The traditional mitigation framework has emphasized design-time decisions: satellites are designed to incorporate disposal capability, passivation, conjunction avoidance, and so on. Real-time mitigation would extend the framework to include operational decisions: when to perform a deorbit burn, how to adjust mission profiles based on current debris conditions, and when to coordinate with other operators on specific actions. The framework remains in its early stages but represents a potentially significant evolution of mitigation practice.

End-to-end environmental analysis, in which mitigation considerations are integrated with broader environmental considerations across the full lifecycle of space activity, is another emerging concept. Conventional treatment has treated mitigation as a specialized topic distinct from broader environmental considerations. The integration of mitigation with launch emissions, atmospheric impacts, astronomical impacts, and other environmental concerns is in its initial stages but is likely to become more prominent as the cumulative environmental footprint of space activity grows.

Practice	Typical Cost Contribution	Implementation Stage
End-of-life propellant reservation	5–10% of launch mass	Mission design
Passivation procedures	Modest design and ops cost	Decommissioning
Conjunction avoidance capability	Operational propellant + systems	Continuous
Casualty risk mitigation	Variable by approach	Mission design
Total estimated mitigation cost	~3–10% of total mission cost	–

Table 10.1. Mitigation Practices and Their Cost Contributions

Chapter Summary

Mitigation, the prevention of new debris generation through systematic design and operational practices, is the foundation of any plausible response to the orbital debris challenge. The principal mitigation practices include end-of-life disposal of satellites and rocket stages, passivation to prevent post-mission explosions, preventing the release of operational debris, conjunction avoidance to prevent accidental collisions, and design measures to reduce the consequences of failures. The international and national standards that codify these practices have evolved over the past three decades and continue to do so, with a trend toward more stringent national requirements and more comprehensive international guidelines. Implementing these standards poses significant compliance challenges, with the most significant issues arising from disposal failures, legacy populations of pre-passivation systems, and new categories of mitigation considerations introduced by large-constellation operations. The frontiers of mitigation practice include cross-jurisdiction coordination, mitigation for in-orbit servicing and active removal, cislunar mitigation, real-time mitigation, and end-to-end environmental analysis. The cumulative effect of mitigation discipline applied across the international space community has produced measurable improvement in the rate of new debris generation per unit of orbital activity. Still, the total rate of debris generation continues to grow due to the rapid increase in overall orbital activity. Mitigation alone, however perfectly implemented, is therefore not sufficient to stabilize the long-term debris environment in supercritical altitude bands; active removal capability, examined in the next chapter, is also required.

Chapter 11. Active Debris Removal and In-Orbit Servicing

Introduction

The cascade analysis in Chapter 4 shows that mitigation alone is insufficient to stabilize the long-term debris environment in the most heavily affected altitude bands of low Earth orbit. The existing population in those bands must be actively reduced, not merely held constant. Active debris removal, the deliberate capture and disposal of derelict satellites and large debris fragments by purpose-built spacecraft, is the technical response to this requirement. Closely related is in-orbit servicing, the deliberate inspection, refueling, repair, or modification of operational satellites by other spacecraft. The two activity classes share substantial technical infrastructure, including capabilities for rendezvous and proximity operations, capture mechanisms, and operational practices, and have developed in parallel over the past two decades.

This chapter examines the technical, operational, regulatory, and economic dimensions of active debris removal and in-orbit servicing. It begins with the technical requirements for these activities and the principal capture and disposal techniques developed. It reviews missions that have been demonstrated or are under development, including ESA's ClearSpace-1, the upcoming PRELUDE mission, the Japanese Astroscale missions, and the U.S. military and commercial servicing demonstrations. It reviews the regulatory and policy framework within which these activities operate, including the legal complications discussed in Chapter 7 and the developing national licensing frameworks. It reviews the economic case for active debris removal and the proposed financing mechanisms. It reviews the prospects for sustained removal capability over the coming decade and its relationship to the broader trajectory of the debris environment.

The Technical Requirements for Active Removal

Active debris removal is technically demanding. The targets are typically tumbling, uncontrolled objects that pose major challenges for capture by another spacecraft. They are typically located in heavily congested orbital regions with dense conjunctions. The removal vehicle must perform a precise rendezvous, capture the target without producing additional fragments, and then either deorbit the captured target or transfer it to a graveyard orbit. Each step of this sequence imposes specific technical requirements that must be met for the mission to succeed.

Rendezvous with a tumbling, uncontrolled target requires the removal vehicle to characterize the target's motion in detail before the approach. The motion can include rapid multi-axis rotation, with rotation rates that may vary over time as the target's residual angular momentum interacts with environmental torques. Optical and radar observations along the approach trajectory provide the data needed to characterize the motion, and the removal vehicle's onboard processing develops the model that supports the capture maneuver. The motion characterization must be precise enough for the capture mechanism to engage the target reliably and fast enough to complete the approach within the operational window.

Capture mechanisms vary widely across the developing approaches. Robotic arms with grippers can grasp specific target features, such as launch interfaces, antennas, or other grappling points. This approach is well-suited to targets with known graspable features that present them to the approaching vehicle. It is less well suited to targets with unknown configurations or to targets whose tumbling motion makes it difficult to engage specific features reliably.

Net capture systems deploy a net that surrounds and constrains the target. The net is typically launched from a launcher on the removal vehicle and closes around the target under the tension of attached lines. This approach is well-suited to large, irregularly shaped targets with limited graspable features. It is less well suited to targets with delicate components that could be damaged by net contact and to operations in environments where the net itself might pose a hazard if it fails to engage properly.

Harpoon systems fire a barbed projectile that embeds in the target structure. The harpoon is connected to the removal vehicle by a tether that enables subsequent control of the captured target. The approach has been demonstrated experimentally and offers the advantage of engaging targets at a distance, before a close approach. The disadvantages include difficulty penetrating various target materials and the risk of producing fragments at the impact site.

Magnetic capture systems use electromagnets or permanent magnets to attach to ferromagnetic features on the target. The approach is limited to targets with ferromagnetic surfaces but offers gentle engagement that minimizes the risk of fragment generation. Astroscale's ELSA-d mission has demonstrated this approach and is being developed for several follow-on missions.

Adhesive capture systems use adhesive surfaces or specialized engagement mechanisms that attach to the target via chemical or mechanical bonding. This approach is in earlier stages of development than the alternatives but offers potential advantages for specific target categories. It has been demonstrated in laboratory testing but not yet in flight.

After capture, the removal vehicle must dispose of the captured target. For targets in low Earth orbit, the typical approach is controlled deorbit; the removal vehicle uses its propulsion to lower the combined satellite orbit until atmospheric drag removes it. This approach requires an extremely large propellant capacity in the removal vehicle, with the required amount increasing in proportion to the target mass and the degree of orbital altitude reduction. For targets in higher orbits, deorbit may not be practical, and disposal may require transfer to a graveyard orbit. The transfer process requires considerable delta-v and may involve multiple orbital maneuvers, each increasing propellant demands.

ClearSpace-1: ESA's Demonstration Mission

The ClearSpace-1 mission, announced by ESA in 2019 and progressing toward a 2028 launch as of mid-2026, is the most prominent active debris removal demonstration in development. The mission was conceived to remove a Vespa upper-stage adapter in low Earth orbit, but in April 2024, ESA retargeted it to the PROBA-1 satellite. The retargeting was prompted by the discovery that the original target had been struck by a small debris fragment, which fragmented it into several pieces and rendered it unsuitable as a removal target.

The PROBA-1 satellite was launched by ESA in 2001 as a small Earth observation satellite at an altitude of approximately 416 miles (670 kilometers). The satellite operated successfully for many years and was retired in 2018. The retirement left the satellite as a defunct object in low Earth orbit, at altitudes where natural decay would not occur within the disposal timeline applicable at the time of launch. PROBA-1 is therefore a representative example of legacy debris from earlier operational eras and provides a meaningful demonstration target.

The ClearSpace-1 spacecraft is a custom-designed removal vehicle built by ClearSpace SA, a Swiss company spun off from the École Polytechnique Fédérale de Lausanne to develop active debris removal capability. The vehicle uses a four-armed robotic capture system that engages the target through controlled contact with the target's external structure. After capture, the vehicle deorbits the combined satellite-target through controlled reentry. The mission is intended to demonstrate operational capability for active debris removal and to establish the technical basis for follow-on missions.

ClearSpace SA has evolved beyond the initial ESA contract to build a broader commercial portfolio that serves both government and commercial customers. The company has secured additional contracts with commercial operators to remove specific satellites at the end of life and to develop specialized servicing capabilities. The company's commercial development illustrates the broader trajectory of active debris removal as an emerging commercial sector rather than solely a governmental capability.

The ClearSpace-1 mission has faced several challenges that have delayed its progress. The April 2024 retargeting from Vespa to PROBA-1 was one such challenge. Funding constraints, technical development complexities, and integration challenges have led to schedule slippage relative to the original plans. The mission's progress through its current development phase will be a critical test of whether the European active debris removal capability can be sustained through the operational deployment phase.

PRELUDE: The Astroscale Approach

The PRELUDE mission, announced by Astroscale in January 2026 with a planned 2027 launch, represents a different approach to active debris removal. Astroscale, a Tokyo-based company, has conducted several debris-removal demonstrations, including the ELSA-d (End-of-Life Services by Astroscale-demonstration) mission, which conducted capture and release demonstrations in 2021-2022. The PRELUDE mission will demonstrate the rendezvous, inspection, and capture of a tumbling, uncontrolled target, a Japanese H-IIA upper stage that has been in orbit since 2009.

The PRELUDE mission is funded primarily by JAXA, the Japanese space agency, with extensive commercial investment from Astroscale. As of 2026, the mission represents the most advanced commercial debris-removal capability under development and is the principal Japanese contribution to the international active debris-removal program. The mission's technical approach includes close inspection of the target before capture, characterization of its tumbling motion, and capture using a specialized mechanism developed by Astroscale.

The mission's relationship to the broader Astroscale portfolio is important context. The company has developed multiple demonstration missions and operational capabilities, including the ELSA-M mission for commercial debris removal services and Lifetime Extension Services for satellites approaching the end of life. The company's approach has been to build a portfolio of related capabilities that share common technical infrastructure while addressing different commercial needs. The PRELUDE mission will demonstrate the most demanding capability, the capture of an uncontrolled tumbling target, and will establish the basis for subsequent commercial removal services.

The Japanese government has been a strong supporter of Astroscale's development, providing both direct funding and broader strategic support for the company's commercial growth. Japan's strategic interest in active debris removal reflects both the country's well-developed commercial space sector (with economic interests in a sustainable orbital environment) and its broader role in international space policy (which has positioned Japan as a leader on debris-related issues). The successful development of Astroscale's capability has been consistent with this strategic positioning.

U.S. Military and Commercial Servicing Demonstrations

The U.S. Department of Defense has developed extensive in-orbit servicing capabilities through both the DARPA Robotic Servicing of Geosynchronous Satellites program and the Northrop Grumman Mission Extension Vehicles. The DARPA program has developed technologies for robotic servicing in geosynchronous orbit, including specialized robotic arms, capture mechanisms, and operational practices for high-altitude servicing. The Northrop Grumman Mission Extension Vehicles, MEV-1 and MEV-2, have demonstrated operational servicing of commercial satellites by docking with target satellites and providing extended station-keeping and maneuvering capabilities.

MEV-1 docked with Intelsat 901 in February 2020, marking the first commercial in-orbit servicing of an operational geostationary satellite. The mission demonstrated the ability to dock with a satellite not designed for docking, to extend operational capability using the servicing vehicle's propulsion system, and to maintain combined operations over extended periods. MEV-2 docked with Intelsat 10-02 in April 2021 and provided similar services. Both MEVs remain in service, demonstrating the operational sustainability of the servicing model.

The MEV approach demonstrates that in-orbit servicing can be commercially viable for high-value assets such as geostationary communications satellites. The economics of the model depend on the value of the extended operational lifetime and the additional revenue the serviced satellite can generate relative to the cost of the servicing operation. For satellites with high residual value but limited propellant, the servicing operation can be highly profitable for both the satellite operator and the servicing provider. The model's applicability to debris removal is more limited, since debris targets do not generate revenue. Still, the technical capabilities developed for the servicing model translate directly to debris removal applications.

The U.S. Space Force has also developed advanced servicing capabilities through programs that include both classified and unclassified components. The specific capabilities are not publicly detailed. Still, the broad pattern includes the development of rendezvous and proximity operations capabilities, inspection capabilities for assessing satellite status, and various servicing capabilities for U.S. military assets. Capability development has been progressing rapidly, and the U.S. military's service capabilities are among the most advanced in the world. Whether the capability will be applied to debris removal in addition to military servicing is an emerging question in U.S. space policy.

The U.S. commercial sector has begun developing an active debris-removal capability through several companies. Astroscale's U.S. subsidiary, Astroscale U.S., has secured contracts for U.S. military servicing applications. Northrop Grumman has expanded its servicing portfolio beyond MEV to include broader robotic servicing. New entrants, including Starfish Space and Orbital Outpost X, have developed servicing concepts that could include debris-removal applications. Commercial development is in its early stages but progressing rapidly, with substantial venture capital investment supporting multiple companies.

The Broader Astroscale Portfolio

Astroscale has emerged as the most prominent international developer of active debris removal capabilities, and its portfolio warrants a more detailed examination. The company was founded in Tokyo in 2013 and has progressively expanded its technical capabilities and commercial reach. The company's mission spans demonstration activities, commercial removal services, and adjacent capabilities, including lifetime extension and lifecycle management.

The ELSA-d mission, launched in March 2021 from Baikonur, demonstrated basic capture-and-release capabilities through a series of operations involving a chaser spacecraft and a target spacecraft, both deployed on the same launch. The chaser was equipped with a magnetic capture mechanism that engaged the target via a docking plate. The mission successfully demonstrated capture and release across multiple operational scenarios, providing the technical basis for subsequent operational missions. The mission encountered operational challenges, including a propulsion system anomaly that limited later phases of the demonstration, but the principal technical objectives were achieved.

The ADRAS-J mission (Active Debris Removal by Astroscale-Japan), launched in February 2024, demonstrated rendezvous and inspection of a real debris target, a Japanese H-IIA upper stage. The mission did not include capture, focusing instead on demonstrating approach, station-keeping, and inspection capabilities. These capabilities were successfully demonstrated in 2024 and 2025, providing detailed observations of the H-IIA target and establishing the operational basis for the follow-on PRELUDE mission, which will attempt to capture it.

The ELSA-M mission, in development for launch in the late 2020s, will provide commercial debris-removal services for OneWeb satellites. The mission will involve capturing failed OneWeb satellites with a dedicated removal vehicle and disposing of them through controlled deorbit. The mission represents the first contracted commercial debris-removal service and provides a model for the broader commercialization of such services.

Astroscale's lifetime extension services represent a distinct commercial application of the company's technical capabilities. The services involve docking with operational satellites at the end of life and providing extended station-keeping capability through the servicing vehicle's propulsion. The model is commercially equivalent to the Mission Extension Vehicle approach demonstrated by Northrop Grumman. However, Astroscale's implementation is intended to address the medium- and lower-mass satellite classes that the MEV model has not addressed.

The company's portfolio reflects the broader trajectory of the active debris removal sector. The capabilities developed for debris removal translate directly to adjacent applications, such as lifetime extension, in-orbit servicing, and mission upgrades. The commercial viability of any individual application may be limited, but integrating multiple applications on a common technical platform yields a viable business model. This pattern is likely to continue to evolve across multiple operators in the coming years.

Selection of Removal Targets

The selection of priority targets for active debris removal is informed by the cascade analysis in Chapter 4 and by the operational considerations of the removal missions. The technical and policy literature has converged on a relatively consistent set of priority targets, with broad agreement across multiple analyses on which specific objects should be addressed first.

The principal selection criteria are the object's contribution to long-term collision risk, the operational accessibility of the object for removal missions, and the political acceptability of the removal operation. The first criterion is technical: large objects in heavily affected altitude bands that would produce many fragments if struck contribute most to long-term risk and provide the greatest benefit if removed. The second criterion is operational: objects reachable by removal missions launched from typical launch sites without prohibitive plane changes are more practical to remove. The third criterion is political: objects whose removal requires the consent or cooperation of multiple states create complications that some removal missions cannot address.

Applying these criteria identifies several priority target categories. Soviet-era SL-16 (Zenit) rocket bodies in sun-synchronous and similar orbits are repeatedly identified as high-priority targets across analyses. These large stages, each weighing approximately 9 metric tonnes (9.9 tons), orbit at altitudes of 700 to 900 kilometers (435 to 559 miles) within the most heavily affected debris band. Approximately 40 such stages remain in orbit, and their removal would help reduce long-term collision risk in their orbital zones.

The SL-16 stages most frequently cited in priority analyses share a common profile. Each is a Zenit-2 second stage with a dry mass of about 8,800 kilograms, an operational propellant capacity that was never fully depleted in earlier missions, and an orbital inclination between 65 and 98 degrees. The top candidates in ESA's and NASA's independent analyses consistently include objects cataloged as 1987-016B, 1988-001B, 1989-033B, and several similar stages from Zenit launches in the late 1980s and early 1990s. Together, these objects have a combined mass of roughly 200 metric tons in the most hazardous altitude band of low Earth orbit. If any one of them were catastrophically fragmented by a collision, the resulting fragment cloud would be comparable in scale to the Iridium-Cosmos event. It would persist for decades at the affected altitude.

Defunct Earth observation and reconnaissance satellites in similar altitude orbits are also priority targets. ENVISAT, discussed in Chapter 10, is the largest single target in this category. Other large legacy Earth observation satellites that have been retired or decommissioned present similar opportunities for removal. The political complications of removing these satellites depend on the launching state and the satellite's operational status; satellites whose launching states are willing to cooperate can be addressed through bilateral arrangements, while those whose launching states are uncooperative pose more difficult diplomatic challenges.

Specific large fragments from past breakup events, particularly the Fengyun-1C and Iridium-Cosmos events, are sometimes identified as priority targets. The argument is that these fragments are the most consequential among the populations affected by those events, and that their removal would reduce the cumulative consequences. The technical and operational challenges of removing fragments differ from those of removing intact satellites, and the priority of fragment removal versus intact satellite removal depends on specific analytical assumptions.

The operational characteristics of specific removal vehicles have also influenced target selection. ClearSpace-1's retargeting from Vespa to PROBA-1 illustrates the practical constraints: the original target was rendered unsuitable by a fragmentation event, and the replacement target had to be selected from the population of available alternatives that the mission's specific capabilities could address. Target selection driven by mission-specific considerations rather than purely by priority analysis is likely to continue, particularly during the demonstration phase of removal missions.

Regulatory and Legal Complications

The legal framework discussed in Chapter 7 poses significant challenges for active debris removal operations. The Outer Space Treaty's principle of state jurisdiction over space objects, combined with the Liability Convention's provisions on liability for damage in space, requires that any removal mission targeting fragments or satellites produced by another state's space activities address several legal questions.

The most fundamental question is consent. Under the prevailing interpretation of the Outer Space Treaty, the launching state retains jurisdiction over its space objects, including debris fragments derived from those objects. Under this interpretation, active removal of such objects without the launching state's consent would violate the treaty's principles. This interpretation has not been formally tested, and no removal mission has yet attempted to address fragments from a non-cooperating launching state. However, the legal opinion among space law scholars is that consent would be required.

The practical implication is that active debris removal missions must either target objects produced by the operator's own state, target objects produced by states that have provided consent, or proceed under specific legal arrangements that address consent. ClearSpace-1's targeting of the PROBA-1 satellite (an ESA-launched object) and Astroscale's PRELUDE mission targeting a JAXA-launched H-IIA upper stage avoid the consent issue by focusing on objects produced by the operator's home agency. Operators that wish to address the highest-priority targets, including Russian SL-16 stages, Chinese fragments from Fengyun-1C, and other objects, face greater legal complications.

The development of frameworks to address these complications is an active area of debris policy. Several specific approaches have been proposed. Bilateral arrangements between launching states and removal operators could address specific removal operations through targeted agreements. Multilateral frameworks could establish standardized consent procedures for operators to follow. Establishing an international coordination body for active debris removal could provide the institutional infrastructure for handling consent at a global scale. None of these approaches has been formally adopted, but discussions are ongoing in COPUOS and other forums.

A second category of legal complications concerns liability for the removal operation itself. If a removal vehicle damages a target during the operation, or if the operation produces additional fragments that damage other objects, the liability for that damage is unclear. The Liability Convention's framework addresses damage caused by space objects but was not designed for the specific scenarios posed by removal operations.

Developing liability frameworks that specifically address removal operations is a key aspect of debris policy.

National licensing frameworks for active debris removal are also being developed. The U.S. FCC has begun addressing removal services through specific guidance, though the framework remains in its early stages. ESA has developed internal procedures for the removal missions it operates and has provided guidance for European commercial operators. The UK Space Industry Act framework includes provisions addressing removal operations, with specific guidance issued by the UK regulator. Japan's framework has been adapted to address Astroscale operations. Jurisdictional variation remains substantial, with consequences for operators seeking to provide services across multiple jurisdictions.

The Economics of Active Removal

The economic case for active debris removal is among the most challenging aspects of the broader debris policy framework. The benefits of removal, including reduced collision risk for the international community, the prevention of additional fragmentation events that would generate thousands of new fragments, and the preservation of operational orbital regions, are diffuse and shared among many operators and states. The costs of removal, including the development of removal vehicles, the operations of removal missions, and the associated regulatory and insurance costs, are borne by specific entities. The misalignment between who pays and who benefits is a textbook commons problem and creates persistent challenges for financing.

The cost of individual removal missions is extremely high. The ClearSpace-1 mission is estimated to cost $113-170 million USD (€100-150 million) across its development and operational phases. The PRELUDE mission's costs have not been fully disclosed publicly, but are likely in a similar range. Follow-on commercial removal missions are expected to be less expensive as the technology matures. Still, the unit cost is unlikely to fall below tens of millions of dollars per mission for the foreseeable future. The total cost of removing the priority targets identified in the cascade analysis, on the order of dozens of large objects, is therefore on the order of billions of US dollars.

The benefits of removal are shared across the international space community. The reduction in collision risk benefits all operators in the affected altitude bands, not just the operators or states that funded the removal. Preventing cascade growth benefits all future users of the affected orbital regimes. The preservation of operational capacity benefits all states with future space ambitions. None of these benefits flows specifically enough to any individual operator or state to justify the full cost of removal under typical commercial logic.

Several mechanisms have been proposed to address the financing challenge. Public funding from major spacefaring states is the most obvious mechanism and underpins ClearSpace-1, PRELUDE, and most other current programs. The model is consistent with other examples of public investment in shared environmental goods, such as climate research and conservation programs. The limitation is that political considerations constrain public funding, and the level of funding may not align with the problem's technical scale.

International financing mechanisms could spread the cost across multiple states. Proposals include dedicated funds administered by UN bodies, contributions tied to launch activity, and other arrangements. None of these proposals has been formally adopted, and the political obstacles to international financing are formidable. Whether any such mechanism will emerge depends on building sufficient political consensus to support it, and that consensus has not yet materialized.

Insurance-based mechanisms could finance the removal of specific high-risk objects through premiums collected from satellite insurance. Several proposals have been developed under this model, with varying degrees of detail. The principal challenge is that the insurance market does not currently price long-term debris risk into premiums in ways that would generate meaningful funding for removal. Pricing long-term risks in insurance markets tends to be conservative for risks that have not yet been demonstrated through actual claims.

User fee mechanisms could collect contributions from operators based on their use of orbital resources, with the proceeds funding the removal of priority targets. The principle is well established in other resource management contexts but has not been applied to orbital activity. Various proposals have been developed, but political obstacles to implementing user fees remain considerable, particularly because the international community lacks the institutional infrastructure to collect them. Whether the user fee approach will emerge as a financing mechanism remains uncertain.

Hybrid mechanisms that combine several of the approaches above are also under discussion. A mix of public seed funding, commercial development, and supporting mechanisms could provide the financing needed for sustained removal capability over the coming decades. The institutional and political development required for such a hybrid approach is extensive, and progress has been slower than the technical development of removal capability.

Sustained Removal Capability

The transition from demonstration missions to sustained removal capability is among the most consequential developments anticipated for the coming decade. The current programs, ClearSpace-1, PRELUDE, and the various supporting missions, are demonstration activities that establish technical capability but operate at scales far below what the cascade analysis suggests is required. Developing sustained capability requires both continued technical maturation and the development of institutional and economic frameworks that can support multi-decade operations.

Technical maturation is progressing reasonably well. The ELSA-d demonstration of 2021-2022, the ADRAS-J inspection of 2024-2025, the ClearSpace-1 launch planned for 2028, and the PRELUDE launch planned for 2027 represent a progression of increasingly capable demonstrations. MEV operations from 2020 onward have demonstrated sustained operational capability for in-orbit servicing, which translates directly into debris removal. The accumulation of operational experience across multiple missions provides the technical basis for next-generation capability development.

Institutional and economic development is progressing more slowly. The legal complications surrounding consent to remove objects from non-cooperating states remain largely unaddressed. Financing mechanisms remain primarily dependent on national space agency budgets rather than on more sustainable mechanisms. International coordination of removal activities is still in its early stages. Each of these areas requires continued development for sustained capability to emerge.

A key question for developing sustained capability is the operational scale that can realistically be achieved. The cascade analysis suggests that several large objects per year would need to be removed from the most heavily affected altitude bands to reduce long-term collision risk effectively. The current demonstration programs operate at scales of one or two missions per agency over several-year periods. Transitioning from this scale to several missions per year would require extensive expansion of both technical capacity and economic resources. Whether the transition will occur depends on factors including the continued maturation of the technology, the development of financing mechanisms, and the willingness of national space programs to commit to sustained removal as a strategic priority.

The relationship between active debris removal and broader space sustainability is increasingly recognized in international policy discussions. The 2024 inclusion of active debris removal as a specific topic in COPUOS discussions, the recent development of national removal strategies in multiple countries, and the growing commercial interest in removal services all reflect a broader recognition that removal capability is essential to long-term space sustainability. Whether this recognition translates into the long-term investment that sustained capability requires will be a central test of debris policy over the coming decade.

In-Orbit Servicing as a Related Capability

In-orbit servicing is a related capability that shares much of the technical infrastructure used for active debris removal. The MEV missions discussed earlier in this chapter have shown that operational servicing of commercial satellites is technically feasible and commercially viable for specific applications. The broader development of servicing capability has implications for debris policy beyond direct removal applications.

The relationship between servicing and removal is multifaceted. Many satellites that would otherwise become debris at the end of their operational lives can have their operational lives extended through servicing, thereby reducing the rate at which new debris is added to the population. Satellites that have failed in ways that prevent normal disposal can sometimes be addressed through servicing operations that restore disposal capability. Satellites that pose specific debris risks due to design or operational issues can sometimes be modified through servicing to reduce those risks.

Developing refueling capability is a particularly important aspect of satellite servicing. Satellites that run out of propellant cannot perform end-of-life disposal maneuvers, and propellant exhaustion has been a significant cause of disposal failures. In-orbit refueling, demonstrated experimentally by several programs and under development for operational deployment, would address this failure mode and improve disposal compliance. The Northrop Grumman Mission Robotic Vehicle, in development for launch later in the 2020s, will incorporate refueling capability and represent one of the most advanced operational refueling demonstrations.

Developing inspection and assessment capability is another important aspect. Satellites that experience anomalies during operation can sometimes be diagnosed and addressed through close inspection by servicing vehicles. This capability is most relevant for high-value assets in geostationary or other high-value orbits, where the cost of detailed servicing is justified by the value of preserving the operational satellite. The capability has proven successful in several missions and is being developed for broader operational use.

The CONFERS organization mentioned in Chapter 8 has been the principal industry body coordinating the development of servicing standards. The standards address technical interoperability (enabling servicing vehicles from one operator to service satellites operated by another), safety practices for proximity operations, and other aspects of operational coordination. The development of standards has been progressive, with the most basic technical compatibility standards published and more advanced operational standards under development.

The Future of Active Debris Removal

The trajectory of active debris removal over the coming decade depends on several developments whose outcomes remain uncertain as of mid-2026. The successful operation of ClearSpace-1, PRELUDE, and other near-term missions will establish the operational viability of the technical approaches. The development of sustained financing will determine whether the capability can scale beyond demonstration. The resolution of legal complications surrounding the removal of foreign objects will determine the achievable scope of removal operations. The integration of removal with the broader regulatory and operational frameworks for space activity will determine the institutional context in which removal operates.

Several specific developments are anticipated. The maturation of commercial removal services will likely create a market with multiple competing operators serving satellite operators, regulatory bodies, and other customers. The development of standardized interfaces, common docking mechanisms, common operational protocols, and common regulatory pathways will reduce barriers to entry and broaden the scope of operations. The integration of removal with insurance markets and commercial financial frameworks will provide additional financing mechanisms. The continued development of international coordination frameworks will address legal complexities and cross-jurisdictional questions.

The challenges to developing sustained capability are substantial. The political obstacles to international financing remain significant. The legal complications of removing foreign objects without consent are not easily resolved. The economic case for removal remains constrained by the diffuse nature of the benefits. The institutional infrastructure needed for sustained operations across multiple jurisdictions and decades has not yet emerged.

The relationship between active removal and mitigation also warrants continued attention. The chapters that follow will examine other aspects of the broader response to the debris challenge, including artificial intelligence and emerging technologies (Chapter 12), economics and insurance (Chapter 13), human spaceflight (Chapter 14), astronomy and the orbital commons (Chapter 15), and the cislunar future (Chapter 16). Active removal is one element of a broader integrated response, and its effective deployment depends on the development of all related elements. Whether the international community can develop an integrated response at the pace and scale suggested by the cascade analysis remains the central question in contemporary debris policy.

Figure 11.1. ClearSpace-1 Capture Sequence (Schematic)

Rendezvous, stationkeeping, robotic capture, and controlled deorbit of PROBA-1

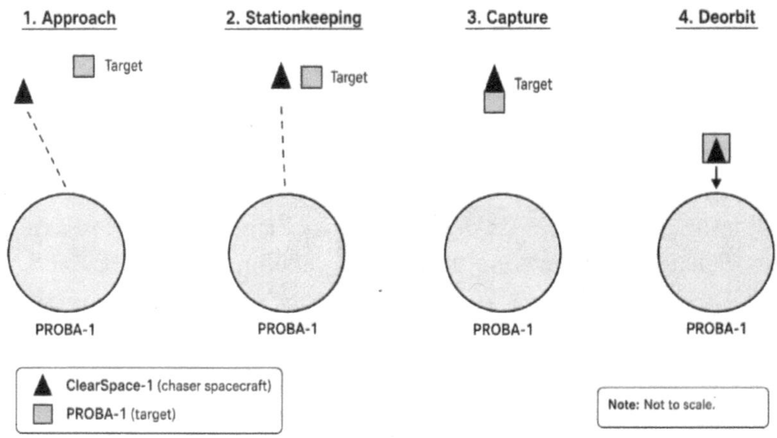

Figure 11.2. Astroscale Mission Timeline: ELSA-d, ADRAS-J, PRELUDE, and ELSA-M

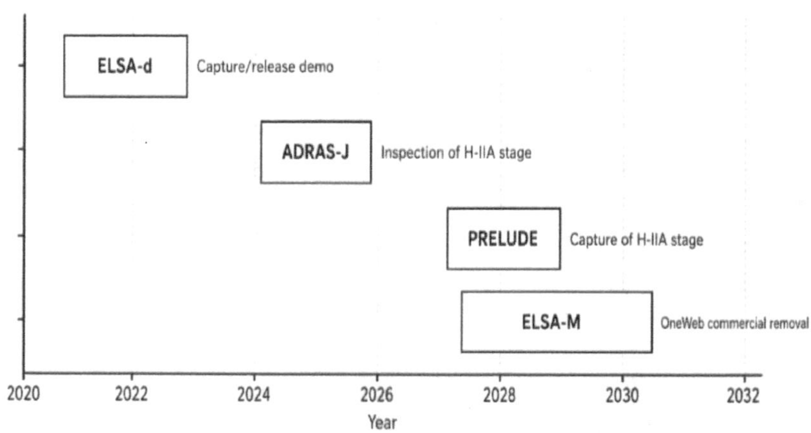

Mission	Operator	Launch	Target	Capability
ELSA-d	Astroscale	March 2021	Magnetic docking plate (synthetic)	Capture/release demo
ADRAS-J	Astroscale / JAXA	Feb 2024	H-IIA upper stage	Rendezvous and inspection
MEV-1 / MEV-2	Northrop Grumman	2019/2020	Intelsat 901 / 10-02	Commercial servicing (GEO)
ClearSpace-1	ClearSpace SA / ESA	2028 (planned)	PROBA-1 (retargeted Apr 2024)	Robotic capture & deorbit
PRELUDE	Astroscale / JAXA	2027 (planned)	H-IIA upper stage	Capture of tumbling target
ELSA-M	Astroscale	2027–2028	OneWeb satellites	Commercial removal service

Table 11.1. Active Debris Removal Demonstrations, 2021–2028

Chapter Summary

Active debris removal, the deliberate capture and disposal of derelict satellites and debris fragments by purpose-built spacecraft, is the technical response to the cascade analysis's conclusion that mitigation alone is insufficient to stabilize the long-term debris environment in supercritical altitude bands. The capability has progressed from concept to demonstration through a series of missions, including ESA's ClearSpace-1 (2028), Astroscale's PRELUDE (2027), and supporting demonstrations such as ELSA-d, ADRAS-J, and the MEV servicing missions. Technical approaches include robotic capture, net systems, harpoon systems, magnetic capture, and adhesive capture, each suited to specific target categories. Target prioritization is informed by the cascade analysis, with Soviet-era SL-16 stages, defunct Earth observation satellites, and specific large fragments identified as the highest priorities. Legal and regulatory complications, particularly the consent issue for removing objects from non-cooperating launching states, remain unaddressed and constrain the achievable scope of operations. The economic case is challenging due to the diffuse nature of the benefits, with public funding from major spacefaring states the principal current financing mechanism. The transition from demonstration to sustained capability over the coming decade will be a central test of debris policy. In-orbit servicing is a related capability that shares much of the technical infrastructure and has additional implications for debris policy through the extension of operational lifetimes, the refueling of satellites for reliable disposal, and the inspection and assessment of operational anomalies. The integrated development of removal and servicing capabilities, combined with ongoing improvements in mitigation, provides the technical basis for the response to the debris challenge as framed by the cascade analysis.

Chapter 12. Artificial Intelligence, Autonomy, and Emerging Technologies

Introduction

The technical response to the orbital debris challenge has been shaped by capabilities that did not exist or that operated at much smaller scales when the principal frameworks were developed. Artificial intelligence and machine learning, autonomous spacecraft operations, and several specific emerging technologies have transformed both the operational practice of debris management and the prospects for expanded capability over the coming decade. This chapter examines these technologies in detail: the application of artificial intelligence to conjunction screening and operational decision support, the development of autonomous spacecraft capabilities for collision avoidance and other operations, the emerging technologies for active removal and tracking, and the broader question of how the integration of these capabilities is reshaping the technical landscape of debris management.

The treatment is necessarily selective. The technologies discussed here are at varying stages of development, with some operationally deployed at scale and others still in research stages. The selection emphasizes technologies whose impact on debris management is already considerable or is expected to become increasingly influential over the coming decade. The chapter concludes with a discussion of how these technologies interact with the broader policy and operational frameworks examined in earlier chapters and the questions they raise for the institutional development of debris management.

Artificial Intelligence in Conjunction Screening

The application of artificial intelligence and machine learning to conjunction screening has progressed enormously over the past several years. The classical approaches to conjunction screening, examined in Chapter 9, involve numerically propagating orbital elements forward in time, identifying pairs that enter defined screening volumes, and computing the probability of collision for each conjunction. The processes are computationally intensive and produce many warnings of varying quality, requiring an exorbitant amount of human attention to identify which warnings warrant an operational response.

Machine learning approaches have been applied at multiple points in this process. Pattern recognition techniques have been used to identify which conjunction warnings are likely to require operational response, allowing operators to focus attention on the most consequential warnings. Reinforcement learning techniques have been used to optimize maneuver planning, identifying maneuver strategies that minimize propellant consumption while achieving required separation. Anomaly detection techniques have been used to identify orbital tracks that may correspond to undetected fragmentation events, supporting earlier identification of new debris populations.

The U.S. Space Force, ESA, and several commercial operators have developed AI-augmented conjunction-screening capabilities. The specific implementations vary, but several common patterns emerge. Machine learning models are typically used to augment rather than replace classical analytical methods, with the AI providing pattern-recognition capabilities that complement the deterministic calculations of classical approaches. The training data for the AI models is typically drawn from operational records of previous conjunctions, with manual review of which warnings warranted maneuvers providing the supervised training signal. The validation of AI capabilities is performed through both retrospective analysis of historical events and prospective evaluation of operational performance.

The principal advantages of AI-augmented screening include the ability to integrate diverse data sources more effectively than classical methods, to identify subtle patterns that might escape rule-based approaches, and to scale to the larger conjunction volumes produced by growing satellite populations. The principal limitations include dependence on training data that may not represent novel conditions, the difficulty of explaining specific decisions in ways operators can verify, and susceptibility to errors when conditions deviate from the training distribution. The deployment of AI in conjunction with screening has therefore proceeded carefully, with extensive human oversight maintained even where AI capabilities have been operationally deployed.

The Defense Advanced Research Projects Agency (DARPA) Hallmark program, conducted in the late 2010s and early 2020s, was an early systematic investigation of AI applications in space situational awareness. The program developed and demonstrated several capabilities for AI-augmented decision support in space operations, including specific applications to conjunction screening. The program's outputs informed subsequent development by both military and commercial operators and provided some of the technical basis for current operational capabilities.

Commercial development of AI-augmented SSA capabilities has expanded rapidly in recent years. LeoLabs, Slingshot Aerospace, and other commercial providers have developed AI capabilities that augment their core tracking and screening services. The capabilities are typically integrated into the operational tools providers offer satellite operators, with AI components providing specific analytical functions that complement the underlying tracking data. This commercial development has been driven by growing demand from satellite operators for more advanced analytical tools, as well as increasing competitive pressure among providers seeking to differentiate their services.

Autonomous Spacecraft Operations

The development of autonomous spacecraft capability has been one of the most consequential technical developments for debris management over the past decade. Spacecraft can independently evaluate conjunction warnings and execute avoidance maneuvers without ground intervention in routine cases, largely reducing the operational burden on ground controllers and improving response speed, particularly for warnings issued shortly before closest approach.

The Starlink constellation has been a particularly visible demonstration of autonomous capability. Each Starlink satellite carries on-board software that evaluates conjunction warnings received through the ground systems, computes the probability of collision and the maneuvers that would resolve identified conflicts, and executes the maneuvers without ground commanding for routine cases. The autonomy is bounded by configuration parameters set by ground operators, including the probability thresholds at which automatic maneuvers are executed and the categories of conjunctions that require human review. The result is a constellation that can respond to thousands of conjunction warnings per year with minimal burden on ground operators. SpaceX's publicly released figures put the 2023 total at approximately 25,000 avoidance maneuvers across the Starlink constellation, rising to approximately 50,000 in 2024 as both the constellation and the surrounding debris environment grew. Translated into a per-satellite figure, this amounts to roughly four to six maneuvers per satellite per year, with the rate varying significantly by orbital shell and proximity to dense conjunction zones. Each maneuver consumes between 0.1 and 0.5 meters per second of delta-V from the satellite's propellant budget, cumulatively reducing the propellant available for end-of-life disposal. The margin is not unlimited, and it narrows as the conjunction rate rises.

The OneWeb, Amazon Leo, and other major commercial constellations have developed similar autonomous capabilities to varying degrees. The specific implementations differ across operators, but the general pattern of autonomous response to routine conjunctions, with human oversight for unusual cases, has become standard practice for leading constellation operations. Smaller operators have generally adopted less autonomous approaches, with more conjunctions requiring human review, but the industry trend is toward greater autonomy. The technical foundations of autonomous conjunction avoidance involve several specific capabilities. On-board orbital propagation, including atmospheric drag modeling and other perturbation effects, must be sufficiently accurate to support the decision-making. Conjunction probability calculations must be performed using the operationally appropriate algorithms with the available data. Maneuver planning must integrate the satellite's operational constraints (propellant availability, attitude control capability, mission requirements) with the safety objective of achieving adequate separation. Each of these capabilities requires substantial computing capacity, software development, and operational verification.

The verification of autonomous systems is particularly challenging. The complexity of the conjunction environment and the diverse circumstances under which decisions must be made create a vast space of possible scenarios, and exhaustive testing is infeasible. The principal verification approaches involve a combination of formal analysis (mathematical proof of specific properties), extensive simulation (testing the system across representative scenarios), and operational monitoring (tracking system behavior in actual operations). Each approach has limitations, and the overall verification confidence depends on the combination of approaches.

The integration of autonomous spacecraft into the broader conjunction-screening infrastructure has introduced new operational dynamics. When two spacecraft both autonomously decide to maneuver in response to the same conjunction, coordination is required to prevent both from maneuvering in ways that produce a new conjunction. Various coordination mechanisms have been developed, including pre-conjunction communication between operators, established protocols for which operator has priority in specific scenarios, and automated negotiation between satellite control systems. The development of robust coordination protocols is an active area of research as autonomous capabilities become more widespread.

AI for Predictive Maintenance and Anomaly Detection

Beyond conjunction screening, AI techniques have been applied to several aspects of spacecraft operations that affect debris generation. Predictive maintenance, which identifies components approaching failure before failure occurs, is one such application. The conventional approach to spacecraft operations has been reactive: components are operated normally until failure occurs, and the consequences are addressed after the fact. AI techniques applied to spacecraft telemetry can identify patterns that precede failures, allowing operators to take preemptive action that may prevent the failure or reduce its consequences.

Battery degradation, propellant system anomalies, and attitude control irregularities are specific failure categories to which AI techniques have been applied to predict. The patterns that precede each failure mode are often subtle and may not be evident in standard telemetry monitoring. Machine learning models trained on historical telemetry from satellites that have experienced specific failure modes can identify similar patterns in operational satellites, supporting earlier intervention.

Anomaly detection, more broadly, is another application. The volume of telemetry from modern satellites, particularly large constellations, is too large for human operators to monitor comprehensively. AI techniques can identify anomalous patterns in the telemetry that warrant human attention, supporting more effective monitoring than would otherwise be feasible. The application has been particularly important for constellation operations where individual operator attention to each satellite is impractical.

The application of AI techniques to spacecraft operations has been progressing substantially across the industry. Major operators, including SpaceX, ESA, NASA, and various commercial entities, have developed and deployed AI-augmented operational capabilities. The capabilities have produced measurable improvements in operational efficiency and reliability. However, the specific quantitative impact is difficult to assess publicly because operators typically do not disclose details of their operational tools.

Machine Learning for Debris Population Modeling

The modeling of long-term debris environment evolution, examined in Chapter 4, has traditionally relied on physical models calibrated against observational data. The models propagate orbital elements forward in time, simulate fragmentation events, and produce projections of the future debris population. Recent work has applied machine learning techniques to augment these classical models, with particular emphasis on capturing patterns in observational data that physical models may not fully represent.

Machine learning approaches have been particularly useful for several specific aspects of debris modeling. The fragment-generation patterns from breakup events depend on the energies and configurations of the colliding objects in ways that are difficult to capture with closed-form models. Machine learning models trained on observed fragment distributions from past events can produce more accurate predictions of fragment distributions for future events. Atmospheric drag modeling for propagation is similarly challenging due to the complex interactions among solar activity, atmospheric composition, and spacecraft characteristics. Machine learning models trained on observational data can produce drag predictions that are more accurate than first-principles models.

The integration of machine learning with physical models in debris simulation is an active area of research. Hybrid approaches that use physical models for the core dynamics and machine learning models for specific aspects that are difficult to model from first principles have produced encouraging results in academic studies. The deployment of these hybrid approaches in operational debris simulation has been more limited, in part due to the verification challenges inherent in operational use.

The application of machine learning to the analysis of observational data has expanded considerably in recent years. Detecting and tracking small debris objects through radar and optical observations presents complex pattern-recognition challenges that machine learning techniques have proven effective at addressing. Modern operational systems for both ground-based and space-based sensors typically include machine learning components for data processing and target identification.

Computer Vision for Debris Tracking and Removal

The use of computer vision in debris-related operations has been important across several specific domains. Detecting small debris from optical sensors involves identifying faint moving objects against a background of stars and other features. Machine learning approaches, particularly deep neural networks for image processing, have improved detection capabilities compared with classical image processing methods. Ground-based optical telescope networks operated by ExoAnalytic and other commercial providers, as well as space-based optical sensors deployed by NorthStar and other operators, rely on machine learning techniques for object detection and tracking.

The application to active debris removal is particularly important. Removal vehicles must characterize the motion of tumbling targets in detail before approach, and this characterization is performed primarily through optical observations along the approach trajectory. Machine learning techniques for processing optical data, identifying specific features on the target, and estimating the target's rotational state have been extensively developed for active debris removal applications. The ADRAS-J inspection of the H-IIA upper stage during 2024–2025 generated valuable operational data for these techniques. It contributed to the advancement of future capabilities planned for the PRELUDE mission.

The application to in-orbit servicing is equally important. Servicing vehicles approach target satellites whose configurations may not be precisely known, and computer vision is the primary means of characterizing the target before specific servicing operations. Mission Extension Vehicle operations have used computer vision extensively for docking with target satellites, and follow-on servicing missions are expected to expand its use.

The development of computer vision capabilities for these applications has benefited greatly from the broader development of machine learning techniques in non-space applications. Many of the algorithmic advances have been adapted from terrestrial computer vision applications, with specific modifications for the space environment. Cross-fertilization between space and non-space applications has been a principal driver of the technical development of AI-related capabilities in space operations.

Emerging Technologies for Active Removal

Beyond the AI and autonomy capabilities discussed above, several emerging technologies are being developed for active debris removal and related applications. Each technology has reached a different stage of development, with some operationally demonstrated and others still in research.

Electrodynamic tether systems harness the interaction between a long conductive tether and Earth's magnetic field to generate drag without consuming propellant. A spacecraft equipped with an electrodynamic tether can gradually deorbit under drag, with the tether converting kinetic energy into electrical energy that is dissipated as resistive heating. The technique is particularly attractive because it operates without propellant, making it suitable for spacecraft that have exhausted their propulsion capability or that were not equipped with significant propulsion. Several demonstration missions have been conducted, including JAXA's HTV-KITE experiment and various academic missions. Still, technical challenges, including tether stability and the management of tether-induced spacecraft dynamics, have limited the operational deployment of the technique.

Drag enhancement devices, such as deployable sails and inflatable structures, increase a spacecraft's cross-sectional area, thereby increasing atmospheric drag at the end of the mission. The principle is straightforward: a satellite that doubles its cross-section experiences twice the drag force at any given altitude. However, operational implementation requires reliable deployment mechanisms and structural designs that preserve the increased area throughout the deorbit period. Several commercial products are available, and the technology has been deployed on various small satellites. The application is most relevant to satellites operating at altitudes where natural decay can be accelerated by drag enhancement to meet applicable disposal timelines.

Laser-based debris removal uses high-power lasers, either ground-based or space-based, to ablate material from debris fragments, producing small thrust forces that gradually modify their orbits. The technique can either remove fragments by deorbiting them through accumulated thrust over many laser passes or modify their orbits to reduce specific collision risks. The technology has been demonstrated experimentally and extensively analyzed theoretically, but its deployment remains constrained by technical, operational, and policy considerations. The technical challenges include developing laser systems with adequate power and beam quality; the operational challenges include targeting and tracking small fragments; and the policy challenges include the dual-use considerations of high-power laser systems with potential offensive applications.

Plasma-based deorbit systems use plasma generation to produce small drag-like forces on satellites. The principle is similar to electrodynamic tethers but relies on different physical mechanisms. Several demonstration missions have been conducted, and the technology offers advantages over tethers, including simplified deployment, yet operational deployment has been limited.

Foam injection systems deliberately create foam structures around satellites at the end of their life, increasing their cross-sectional area to accelerate atmospheric drag. The technology has been extensively studied but has not been operationally demonstrated. The principal challenges are the long-term stability of the foam structures and the operational complications of foam injection in the space environment.

Next-Generation Tracking Sensors

Next-generation tracking sensors are advancing across multiple programs and are likely to expand the global tracking infrastructure dramatically over the coming decade. Several specific developments warrant attention.

The U.S. Space Force is developing additional ground-based phased-array radars to greatly expand tracking capability beyond the current Space Fence at Kwajalein. The specific systems and locations have not been fully disclosed publicly, but the broad pattern includes additional radars at multiple sites to provide more comprehensive coverage. The systems are expected to provide both a larger tracking volume and improved tracking precision relative to current capabilities.

Multiple programs have deployed space-based optical sensors and are likely to expand significantly over the coming decade. NorthStar Earth and Space's first three operational satellites entered service in 2025, providing observations from low Earth orbit that complement ground-based sensors. The U.S. Space Force operates the Geosynchronous Space Situational Awareness Program, which provides space-based observations of objects in the geostationary belt and other high-altitude orbits. Additional commercial operators, including LeoLabs, are developing space-based optical capabilities. Together, these developments are expected to expand coverage in high-altitude regions where ground-based observations are limited.

Space-based radar sensors, though less developed than optical sensors, are also advancing through several research and development programs. The technical challenges of space-based radar include substantial power requirements and the complexities of operating high-power transmitters in space. Several proposed systems address these challenges through specific designs, and operational deployment is expected within the next decade.

Specialized sensors for specific debris populations are also being developed. Several operators have demonstrated lidar systems for high-precision tracking of nearby objects. Bistatic radar systems that combine separated transmitters and receivers offer capabilities for certain debris populations. Other specialized sensor concepts are under development for niche applications.

Integrating next-generation sensors with the broader tracking infrastructure will require extensive development of data processing and integration capabilities. The data volumes produced by next-generation sensors will exceed the capacity of current data processing systems, necessitating continued expansion. Integrating data from multiple sources, including government and commercial, ground-based and space-based, radar and optical, will require improved data fusion approaches to produce coherent operational products from these diverse inputs.

Quantum Technologies and Other Frontiers

Quantum technologies have attracted increasing research attention, including applications to space situational awareness. Quantum sensing techniques offer potential advantages for precise measurement of position and velocity, which could improve tracking capabilities. Quantum communication techniques offer secure communication that could be relevant for sensitive space operations. The application of quantum technologies to space operations is in its early stages, with most current work in research rather than operational deployment.

Cold-atom interferometry, a quantum sensing technique, has been validated for ground-based applications, including geodesy and inertial measurement. Space-based sensing for SSA is theoretically promising but has not been operationally proven. Several research programs are exploring this application, and operational demonstrations could occur within the next decade.

Quantum communication systems for satellite-to-ground links have been tested by several programs, including the Chinese Micius satellite, which conducted quantum key distribution experiments in 2017, as well as subsequent missions by other operators. The application to debris management is limited, but could become relevant if quantum communication is deployed operationally for space situational awareness data sharing.

Other emerging technologies that may become relevant to debris management include advanced propulsion systems (such as electric propulsion, which may enable more efficient removal operations), additive manufacturing in space (which may enable in-orbit construction of capabilities, including removal vehicles), and various biotechnology applications (which may play specific roles in long-term space operations). Each of these areas is under development, and the operational impact on debris management remains uncertain.

The Integration Challenge

The development of multiple emerging technologies poses significant integration challenges. The current operational environment comprises diverse capabilities developed by different operators using different technical approaches and standards. Integrating these capabilities into coherent operational frameworks requires substantial coordination across multiple dimensions.

Technical integration involves developing interoperability standards that enable capabilities from different operators to work together effectively. The CONFERS organization has developed interoperability standards for in-orbit servicing. Various organizations have developed standards for tracking data exchange. Continued development of integration standards is essential for the broader deployment of advanced capabilities.

Operational integration requires coordinating practices among operators with diverse organizational cultures, regulatory environments, and operational priorities. Commercial operators have established bilateral coordination arrangements, but broader integration of operational practices across the international community remains in its early stages. The Space Safety Coalition and other industry organizations have developed best practices that facilitate operational integration, yet the voluntary nature of these arrangements limits their effectiveness.

Regulatory integration involves coordinating regulatory frameworks across multiple legal systems to support the deployment of integrated capabilities. The variation in national frameworks discussed in Chapter 8 imposes compliance costs on operators that must satisfy multiple regulators. It also creates risks of regulatory arbitrage, in which operators select regulatory environments based on permissive regulation rather than substantive considerations. The development of mutual recognition arrangements and other integration mechanisms is underway, but has been slow.

The institutional integration of the various aspects of debris management also remains in development. The traditional separation of activities into specialized domains, including mitigation, removal, tracking, and conjunction screening, has been useful for organizing technical and policy work. Still, integrating these activities into coherent operational frameworks is essential for effective management of an increasingly complex environment. Developing integrated frameworks is one of the principal challenges of contemporary debris policy.

Implications for the Workforce

The development of AI and autonomous capabilities has profound implications for the workforce required to operate the global debris management infrastructure. The traditional approach has relied on extensive human attention to individual conjunction warnings, manual review of operational decisions, and significant ground controller involvement in spacecraft operations. The transition to AI-augmented and autonomous systems shifts workforce requirements toward more analytical and oversight functions and away from routine operational tasks.

The transition has been gradual rather than abrupt. The major commercial operators have developed advanced AI and software engineering capabilities while maintaining traditional operational expertise. The integration of these capabilities has produced organizations that can support both routine autonomous operations and the human expertise required for unusual circumstances. This pattern is likely to continue across the industry.

Training and developing personnel for the emerging operational environment is a monumental challenge. Traditional educational pipelines for space operations have emphasized aerospace engineering and astrodynamics; the emerging environment requires advanced capabilities in software engineering, machine learning, and data science alongside the traditional disciplines. The space industry competes with broader technology sectors for personnel with the relevant skills, creating corresponding challenges for recruitment and retention.

The implications for workforce development extend beyond the technical disciplines. The operations of large constellations and the integration of complex AI systems require management capabilities that have not traditionally been emphasized in space operations. Developing operational practices that effectively integrate AI with human decision-making, maintain organizational culture across distributed operations, and support continuous learning and improvement requires management approaches that the industry is still developing.

The Role of Open Source and Collaborative Development

The development of AI and emerging technologies for debris management has relied heavily on open-source and collaborative efforts. Several open-source software projects, academic research programs, and industry consortia have contributed to the available capabilities. This collaborative pattern has accelerated capability development compared with closed, proprietary models, but it also creates challenges for verification, security, and accountability.

Multiple academic and government programs have developed open-source software for orbital propagation, conjunction analysis, and related functions. The Astropy library, GMAT (General Mission Analysis Tool) developed by NASA, and OREKIT, developed by various European institutions, along with other tools, provide open implementations of standard algorithms that researchers and operators can build upon. The availability of these tools has significantly reduced barriers to entry for new operators and researchers and has accelerated the development of specialized capabilities.

Academic research programs have made significant contributions to the development of AI and other emerging technologies. Universities, including the Massachusetts Institute of Technology, the University of Texas at Austin, the École Polytechnique Fédérale de Lausanne, the

University of Southampton, and many others, run impressive research programs in space situational awareness and debris management. These programs have produced specific technical advances and trained personnel who later move into operational positions.

Industry consortia have facilitated coordination among operators and researchers. The Space Safety Coalition, CONFERS, the Consortium for Earth Energy Imbalance, and other organizations provide forums for the coordinated development of best practices and standards. The voluntary nature of these consortia has both advantages (flexibility, broad participation) and limitations (limited enforcement, dependence on participant willingness).

The relationship among open-source, academic, and proprietary commercial development has been complex. Many commercial operators have built proprietary capabilities on open-source foundations, contributing improvements back to open-source projects in some cases and maintaining proprietary advantages in others. This balance has enabled rapid capability development but has also created challenges for the broader community in tracking and verifying the capabilities operators deploy.

Implications for Policy and Governance

The development of AI and emerging technologies has far-reaching implications for the policy and governance of space activities. Traditional regulatory frameworks were designed for human-mediated operations, with explicit decisions and clear accountability for those decisions. The transition to AI-augmented and autonomous operations creates regulatory challenges that must be addressed.

Accountability for AI-driven decisions is a specific challenge. When an autonomous spacecraft executes an avoidance maneuver based on its onboard software's evaluation, responsibility for the decision is shared among the operator who deployed the software, the developer who wrote it, the regulatory body that authorized its use, and other parties. Determining liability when a maneuver produces unintended consequences may be unclear in ways traditional regulatory frameworks have not addressed. Developing accountability frameworks for autonomous operations is an active area of policy work.

Verifying AI capabilities for regulatory purposes is another challenge. Certification frameworks for traditional spacecraft systems were based on detailed engineering analysis and testing of specific designs. Verifying AI capabilities, particularly machine learning systems whose behavior depends on training data, requires different approaches that the regulatory community is still developing. The current approach has generally involved a combination of traditional certification of the deterministic components, simulation-based validation of the AI components, and operational monitoring with human oversight, but the framework remains in development.

The dual-use considerations of advanced capabilities are particularly significant. Capabilities developed for debris removal, in-orbit servicing, and similar applications overlap considerably with those relevant to offensive military operations. Maneuvering one spacecraft to capture another can be used either to remove a derelict object or to interfere with an operational satellite. Regulatory frameworks have generally addressed these considerations through licensing arrangements and operational practices, but their integration with broader arms control and security frameworks remains under development.

The international coordination of deploying advanced capabilities is the final challenge. Variations in national approaches to AI and emerging technologies create risks of diverging operational practices that could complicate the broader international management of the orbital environment. The development of common standards and practices through international forums has progressed, but at a slower pace than the development of underlying capabilities. Continued development of coordination mechanisms is essential to support the effective deployment of advanced capabilities.

Future Trajectories

The trajectory of AI and emerging technologies in debris management over the coming decade will be shaped by several factors. Ongoing advances in underlying AI capabilities, driven by broader technological progress, will provide increasingly capable building blocks for space-specific applications. Growing operational demand from constellation operators and other users will support investment in capability development. Evolving policy and regulatory frameworks will shape the deployment of capabilities in operational environments.

Several specific trajectories are anticipated. Capabilities for autonomous spacecraft operations will continue to expand, with more sophisticated decision-making, broader categories of autonomous response, and greater integration across operators. Similarly, capabilities for AI-augmented decision support in ground operations will expand, with more sophisticated analytical tools, better integration of diverse data sources, and greater support for complex operational decisions. The deployment of next-generation tracking sensors will significantly expand the empirical foundation on which all of these capabilities operate.

Future development faces challenges in integration, the workforce, and policy. None of these challenges has an obvious solution, and developing effective approaches will require sustained attention from operators, regulators, researchers, and policymakers. The pace at which these challenges are addressed will greatly affect how quickly AI and emerging technologies translate from technical capability to operational impact in the debris environment.

The deeper question is whether the advanced capabilities will be deployed at a scale and pace sufficient to address the debris challenge meaningfully. The cascade analysis identifies specific quantitative requirements for stabilizing the long-term debris environment. The current trajectory of capability development is encouraging, but the gap between current capabilities and requirements remains considerable. Whether this gap can be closed through the continued development of AI and emerging technologies, integrated with the policy and operational frameworks examined in earlier chapters, is one of the central questions for the trajectory of debris management.

Chapter Summary

Artificial intelligence, autonomous spacecraft operations, and several emerging technologies have transformed debris management practices and the prospects for significantly expanded capability over the coming decade. AI applications, in conjunction with screening, have enhanced operators' analytical capabilities, with both military and commercial implementations contributing to this improvement. Autonomous spacecraft operations have been particularly important for large constellations, with Starlink's autonomous conjunction avoidance capability serving as the most prominent example. AI techniques have been applied to predictive maintenance, anomaly detection, debris population modeling, and computer vision for tracking and removal. Emerging technologies for active removal, including electrodynamic tethers, drag enhancement devices, laser-based removal, and others, are at varying stages of development. Next-generation tracking sensors will expand the global infrastructure over the coming decade. Integrating these capabilities into coherent operational frameworks faces enormous challenges across technical, operational, regulatory, and institutional dimensions. The implications for the workforce, policy and governance frameworks, and the international coordination of debris management are extensive. Whether advanced capabilities will be deployed at scale and at a pace sufficient to address the debris challenge meaningfully depends on continued development of capabilities, the development of supporting policy and regulatory frameworks, and the integration of these elements into a coherent response to the cascade dynamics now present in the orbital environment.

Chapter 13. Economics, Insurance, and Sustainability Ratings

Introduction

The orbital debris challenge is fundamentally an economic problem as well as a technical one. The cascade analysis in Chapter 4 establishes the technical conditions for stabilizing the debris environment, but achieving those conditions depends on individual actors' willingness to bear costs whose benefits accrue broadly to the international community. The economic structure of space activity, the costs of mitigation and removal, the pricing of risk in insurance markets, the financing of long-term sustainability investments, and the alignment of commercial incentives with collective interests shape the practical implementation of debris policy as much as the technical capabilities discussed in earlier chapters.

This chapter examines the economic dimensions of orbital debris management. It begins with the broader economic context of space activity, establishing the industry's scale and structure within which debris-related decisions are made. It examines the costs of debris mitigation and how they are distributed. It examines the space insurance market and its role in pricing debris-related risk. It examines the emerging Space Sustainability Rating program and similar efforts to inform commercial decisions about sustainability practices. It examines the broader question of the common-pool character of orbital debris and the economic mechanisms proposed to address common-pool problems in this context. And it examines the developing financing mechanisms for active debris removal and other long-term sustainability investments. The treatment is necessarily integrative, drawing on the technical and policy material from earlier chapters to examine the economic structures within which all of those elements operate.

The Economic Scale of Space Activity

The global space economy has grown over the past decade, driven primarily by the expansion of commercial activity in low Earth orbit. Estimates of the size of the space economy vary by methodology, but recent industry surveys consistently place it at approximately $570-$630 billion annually as of 2024-2025. The projected growth trajectory points toward $1 trillion or more by the early 2030s, driven by continued constellation deployment, the expansion of commercial in-orbit services, and growth in downstream applications, including communications, Earth observation, and navigation services.

The figure is worth breaking down because it is sometimes misread as representing the revenues of space companies alone. Commercial satellite services, primarily telecommunications, Earth observation, and satellite broadband, account for roughly $330 billion of the total. Satellite manufacturing and launch services account for approximately $60 billion. Government space budgets globally, including NASA, ESA, and the national programs of China, Russia, India, Japan, and other spacefaring states, account for roughly $120 billion. Ground equipment, including the hardware that receives and processes satellite signals, accounts for the remaining $120 billion or so. The vast majority of this economic activity depends, directly or indirectly, on access to orbital space. This dependence gives the debris problem its economic urgency.

The structure of the industry has shifted toward commercial operations in recent years. Commercial activity now accounts for most orbital launches, operational satellites, and a large share of total space-related employment and economic activity. The shift has been driven by the success of commercial launch providers, particularly SpaceX, which have drastically reduced launch costs through reusability and other innovations; by the deployment of large commercial constellations; and by the growth of commercial Earth observation, communications, and other downstream services.

The economic structure of the satellite manufacturing and operations sectors has implications for debris policy. Principal commercial constellations operate at scales that yield strong economies of scale in satellite production and launch, with per-satellite costs far lower than those of traditional bespoke satellite missions. These lower costs have enabled the deployment of constellations at scales that would have been economically infeasible under traditional cost structures. The same economics that enable these constellations also shape debris-related decisions. The cost of additional disposal capability per satellite is small relative to the satellite's total cost, but the aggregate cost across thousands of satellites can be enormous.

The geographic distribution of space economic activity has also shifted. The United States retains the largest share of the global space economy, driven by commercial leadership, government spending, and substantial private investment. Europe, China, Japan, India, and other major spacefaring economies have continued to grow, though at varying rates. The emergence of new commercial space hubs, such as Luxembourg, the United Arab Emirates, Australia, and the United Kingdom, has diversified the geographic distribution and produced new patterns of regulatory competition that affect debris-related decisions.

Investment patterns in the space sector have fueled significant growth among commercial space companies. Over the past several years, venture capital and private equity firms have invested approximately $10-15 billion annually in the commercial space sector, supporting the development of commercial constellations, in-orbit servicing companies, active debris removal firms, and related technology providers. These patterns reflect confidence in the commercial viability of the broader space sector but also create pressure for rapid deployment, complicating sustainability considerations.

The Costs of Mitigation

The implementation of debris mitigation imposes specific costs on satellite operators that vary by satellite design and operational environment. Several cost categories warrant particular attention.

End-of-life disposal capability typically requires reserving propellant throughout the mission, with the required amount scaling with mission altitude and the chosen disposal approach. For low Earth orbit satellites disposed of via controlled reentry, the propellant requirement is typically

5 to 10 percent of the satellite's launch mass. The cost implications include both the direct cost of the additional propellant and the opportunity cost of mass that could otherwise be allocated to operational capabilities. For constellation operators that produce satellites at scales of thousands per year, the aggregate cost of disposal propellant is considerable. However, the per-satellite cost is small relative to the total cost of satellites.

Passivation requirements add modest costs to spacecraft design and operation. Modern designs typically incorporate passivation as a baseline feature, with the additional cost integrated into the broader satellite design rather than identified separately. For satellites that incorporate passivation as part of comprehensive end-of-life procedures, the marginal cost is generally small. For satellites not otherwise planned for passivation, the additional cost can be significant, particularly for smaller satellites, where the operational complexity of passivation procedures may be substantial relative to baseline operations.

Conjunction avoidance capability requires both propulsion to execute maneuvers and operational systems to receive and respond to conjunction warnings. The propulsion requirement is integrated with the satellite's broader propulsion needs (station-keeping, attitude control, disposal) and is generally not identified as a separate cost. The operational systems, including ground control infrastructure, software for evaluating conjunction warnings, and autonomous onboard capability for routine response, represent large investments for leading constellation operators and incremental costs for smaller operators.

Casualty risk mitigation, achieved through controlled reentry capability or design-for-demise, entails specific costs that vary by approach. Controlled reentry capability requires propellant reserve, as discussed above, and additional reliability investments in the propulsion and attitude control systems. Design-for-demise requires specific material and structural choices that may entail higher costs than alternative designs. For most modern satellites, one or both of these approaches have been incorporated as baseline features, with the costs integrated into the broader satellite design.

The aggregate cost of mitigation across the satellite industry is difficult to estimate precisely because these costs are typically integrated into broader satellite design and operational costs rather than identified separately. Industry surveys and academic analyses have produced estimates ranging from a few percent to approximately ten percent of total mission cost for typical low Earth orbit satellites. The variation reflects both methodological and substantive differences across satellite types and operational approaches.

The distribution of mitigation costs across operators is uneven. Major constellation operators have invested heavily in mitigation capabilities, driven by both regulatory requirements and the operational benefits of reliable mitigation in their dense conjunction environments. Smaller operators with single satellites or small fleets may face proportionally higher per-satellite costs because they cannot achieve the economies of scale that constellation operators enjoy. This distribution has implications for the broader competitive dynamics in the satellite industry, with some advocates arguing that mitigation requirements disproportionately affect smaller operators. In contrast, others argue that the requirements appropriately reflect the externalities that all operators impose on the orbital environment.

Space Insurance Markets

The space insurance market is a principal mechanism for managing the financial risks of space activity. The market provides coverage for several distinct categories of risk: launch failures (typically covered by specific launch insurance), in-orbit failures during the operational mission (typically covered by in-orbit insurance), and third-party liability (typically covered by liability insurance). Each category has distinct market dynamics and pricing structures that influence the broader management of debris-related risks.

The global space insurance market remains relatively small in absolute terms, with annual premiums estimated at approximately $400–$500 million as of 2024–2025. The industry has historically been volatile, alternating between periods of profitability and severe financial losses driven by space-related incidents. Events such as the 2009 Iridium-Cosmos collision, along with multiple major launch failures over the years, generated extensive insurance losses that influenced market pricing, underwriting practices, and overall insurance capacity.

Launch insurance prices have generally declined over the past decade, reflecting both improvements in launch reliability (particularly through the operational maturation of the Falcon 9) and broader competitive dynamics in the launch market. Typical launch insurance premiums now range from 1 to 3 percent of the insured value, depending on the launch vehicle and specific mission characteristics. This decline has helped enable the deployment of larger satellite missions and the broader expansion of commercial activity.

In-orbit insurance prices have been more variable, reflecting the broader risk profiles of in-orbit operations. Premiums for typical commercial geostationary satellites range from 0.5 to 2 percent of insured value per year, with higher rates for satellites in more heavily affected debris environments. Pricing reflects both the actual claims experience of in-orbit operations and underwriters' assessments of debris-related risks. The recent growth of constellation operations has raised new questions about insurance pricing for satellites in dense conjunction environments, with some underwriters expressing concern about the long-term claims experience in these new operational settings.

Third-party liability insurance addresses the risk of damage to third parties arising from space activities. Most major jurisdictions require this coverage under national regulatory frameworks, though specific requirements vary by country. The U.S. requires third-party liability insurance for commercial launches, with coverage limits typically in the hundreds of millions of dollars. ESA member states have similar requirements, though the specific coverage varies. Premiums for third-party liability insurance are typically modest relative to launch insurance and reflect the limited claims experience under existing frameworks.

The pricing of debris-related risks in insurance markets has become an increasingly important topic in recent years. The conventional approach has been to incorporate debris-related risks into broader operational risk pricing, with limited explicit pricing of long-term debris exposure. Several developments have begun to shift this approach. Some underwriters have begun differentiating pricing based on operators' debris-mitigation practices, offering premium reductions for satellites with strong mitigation discipline and additional charges for those with weaker mitigation. The Space Sustainability Rating program, examined below, has begun to be used by some underwriters as one input to their pricing decisions. Various proposals for more systematic incorporation of debris-related risks into insurance pricing have been developed but not yet broadly implemented.

The capacity of the insurance market to handle large debris-related events is a principal question for the market's future development. The market has not been tested by a cascade event or by widespread debris-related claims. Whether the existing capacity would be adequate to handle the financial consequences of such an event, or whether additional capacity development would be required, remains an open question that has been the subject of industry analysis. Several proposals for developing catastrophic risk capacity, including reinsurance arrangements specific to space debris and possible governmental backstops for catastrophic events, have been developed but not yet implemented.

The Space Sustainability Rating Program

The Space Sustainability Rating (SSR) program, initiated by the World Economic Forum and now operated by the École Polytechnique Fédérale de Lausanne in Switzerland, provides a structured framework for evaluating and disclosing the sustainability characteristics of space missions. The program assigns ratings based on criteria such as debris-mitigation practices, conjunction-avoidance capabilities, operational transparency, and adherence to international guidelines.

The program operates voluntarily. Satellite operators submit information about their missions for evaluation, and the program issues ratings based on detailed criteria developed through multi-stakeholder consultation. The ratings provide a comparative framework that operators can use to assess their practices against industry norms and that other parties, including customers, investors, and regulators, can use to inform their decisions about specific operators.

The rating criteria are designed to capture multiple dimensions of sustainability practice. The debris-mitigation criteria include specific assessments of disposal practices, passivation, conjunction avoidance, and operational debris management. The transparency criteria include assessments of the operator's data-sharing practices, reporting on operational events, and engagement with the broader space community. The compliance criteria include assessments of adherence to applicable national and international standards. The criteria have been progressively refined based on operational experience and stakeholder feedback.

Participation in the program has been growing. Several major commercial operators have submitted their missions for rating, demonstrating their willingness to undergo structured sustainability evaluation. Participation has been concentrated among operators with strong sustainability practices, who have used the ratings to demonstrate their commitment to operational discipline. The program's challenge has been to expand participation beyond the leading operators to include a broader range of the satellite industry.

The use of the ratings by other parties has been growing, but remains in its early stages. Some insurance underwriters have begun using ratings as inputs to their pricing decisions, adjusting premiums based on rating outcomes. Some institutional investors have begun using the ratings in their environmental, social, and governance (ESG) assessments of space companies. Some satellite customers have begun using the ratings to select satellite operators for specific services. The expansion of these uses, with corresponding commercial impact, is one of the program's developing trajectories.

The SSR's relationship to the broader regulatory framework has been the subject of substantial discussion. The program is voluntary and operates outside formal regulatory structures, yet it provides information that complements regulatory compliance. Some observers argue that the program could become a de facto standard that influences regulatory development. In contrast, others argue that it should remain separate from regulatory frameworks to preserve its voluntary character. The specific trajectory of this relationship will depend on developments in both the program and the broader regulatory environment.

Other sustainability rating efforts have emerged alongside the SSR. Commercial and academic programs have developed evaluation frameworks for specific aspects of space sustainability. The Space Industry Indemnity, the Outer Space Sustainability Index, and several other programs offer alternative or complementary evaluation approaches. The proliferation of rating frameworks brings both advantages (multiple perspectives, competitive innovation) and challenges (potential confusion, varying methodologies). The development of common standards or mutual recognition arrangements among rating frameworks is an emerging aspect of the broader sustainability evaluation ecosystem.

The Commons Problem and Its Economic Structure

The orbital debris challenge has the structure of a classic commons problem. The orbital environment is a shared resource that no individual operator owns, that no individual operator can be excluded from using, and that degrades through individual use, with the cumulative effects affecting all users. This structure creates persistent incentives for individual operators to externalize the costs of their activities onto the broader community of users, and the resulting collective outcome may be worse than what coordinated action could achieve.

The economic literature on commons problems identifies several mechanisms that can address the structure. Property rights, which assign specific use rights that can be traded among operators, can internalize externalities by making individual operators responsible for the consequences of their actions. The assignment of orbital slots in geostationary orbit through ITU coordination is one example of property rights in space. However, these rights are typically assigned to states rather than individual operators, and the mechanism does not directly address debris-related externalities.

Government regulation, the imposition of binding rules that limit degradation of the commons, can address the structure when the government has the authority and capability to enforce them. The national regulatory frameworks examined in Chapter 8 are examples of government regulation, but the international character of orbital activity limits the effectiveness of any single national framework. The international guidelines examined in Chapter 7 represent a partial response, but their voluntary nature limits their capacity for direct enforcement.

User fees, which impose charges for use of the commons to fund resource preservation, can provide both incentive effects (discouraging wasteful use) and revenue effects (financing preservation activities). Various proposals for user fees in the orbital context have been developed, including launch fees, satellite operation fees, and fees tied to specific debris-generating activities. None of these proposals has been broadly implemented, and the political obstacles to international fee mechanisms are immense.

Cooperative arrangements, the voluntary coordination of operators to address common problems, can be effective when participants' interests are aligned, and the institutional infrastructure supports cooperation. Bilateral coordination among major commercial constellation operators exemplifies a cooperative arrangement, and the broader Space Safety Coalition framework provides additional cooperative infrastructure. The voluntary nature of these arrangements limits their effectiveness for actors who choose not to participate.

The combination of these mechanisms in the orbital context has evolved, but this specific combination has not produced a comprehensive response to the common characteristics of the debris challenge. The continued evolution of regulatory frameworks, the development of new mechanisms for international cooperation, and the maturation of voluntary frameworks all contribute to the broader response, yet the gap between the current response and what would be required to address the commons character fully remains considerable.

Externality Pricing and Internalization

A specific challenge in addressing the commons character is pricing externalities, the costs that individual operators impose on the broader community through their debris-generating activities. Conventional economic theory holds that setting the cost equal to the externality for the responsible party can internalize it and align individual incentives with collective interests. Applying this theory to the debris context faces several complex challenges.

Measuring debris-related externalities is inherently difficult. The cost imposed on the international community by a specific debris-generating event depends on the event's long-term consequences, which, in turn, depend on the future evolution of the debris environment, the operational practices of future users, and many other uncertain factors. The development of methodologies for estimating these costs has been an active area of academic and policy research, with various approaches yielding estimates that vary significantly with methodological assumptions.

Attributing externalities to specific operators is similarly challenging. Many debris-generating events involve multiple actors: the operator of a fragmenting satellite, the operator of any colliding object, the launching states for both, and the operators of any subsequently affected satellites. Allocating responsibility for the resulting externalities among these actors is not straightforward, and different allocation approaches yield vastly different conclusions about who should bear the costs.

Implementing externality pricing within operational frameworks faces additional challenges. Collecting fees from operators based on their externality contributions requires institutional infrastructure that does not yet exist. The political consensus needed for international agreement on fee structures has not yet developed. Integrating externality pricing with the existing regulatory and insurance frameworks would require extensive restructuring of the broader space governance ecosystem.

Despite these challenges, the conceptual framework of externality pricing has influenced policy discussions in several important ways. Externality considerations have informed arguments for tighter mitigation requirements, including the U.S. five-year disposal rule and similar developments in other jurisdictions. Similarly, arguments for more systematic incorporation of debris-related risks into insurance pricing share the same motivations. The development of sustainability ratings and similar frameworks reflects an attempt to provide proxy mechanisms for externality pricing through reputational and commercial channels.

The future development of more comprehensive externality pricing frameworks remains uncertain. The major obstacles, including measurement, attribution, institutional infrastructure, and political consensus, are unlikely to be resolved quickly. However, the conceptual framework continues to influence policy discussions and may eventually yield more concrete implementation mechanisms. National experimentation with externality-based fee structures, possibly in specific commercial sectors or orbital environments, may provide a path toward broader implementation.

Financing Active Debris Removal

The financing of active debris removal, briefly examined in Chapter 11, is one of the most challenging economic questions in contemporary debris policy. The misalignment between who pays the costs and who receives the benefits creates persistent obstacles to financing at the scale that the cascade analysis suggests is necessary.

The current financing approach relies primarily on public funding from national space agencies. ClearSpace-1 is funded primarily by ESA, with contributions from member states. PRELUDE is funded primarily by JAXA, with commercial co-investment from Astroscale. Various U.S. military service programs are funded through the Department of Defense budget. This pattern produces specific demonstration missions but does not provide the sustained financing that operational removal would require.

Several proposals for alternative financing mechanisms have been developed. Public-private partnerships could combine public seed funding with private commercial development to build more sustainable removal capabilities. The Mission Extension Vehicle approach demonstrated by Northrop Grumman offers a partial example: the development was largely commercial, but the customer base included significant government participation. Expanding similar arrangements specifically for debris removal could provide a financing model.

International financing mechanisms could spread the cost across multiple states. Proposals include dedicated funds administered by UN bodies, contributions tied to launch activity, and other arrangements. The political obstacles to international financing remain significant, but the continued development of the international space governance framework may eventually provide the institutional infrastructure these mechanisms require.

Insurance-linked financing mechanisms could leverage the existing insurance market to support removal financing. Specific structures could include catastrophic risk pools that fund the removal of high-priority targets to reduce the broader insurance market's exposure to catastrophic risk, premium structures that incorporate removal financing into operating costs, and mutual insurance arrangements among operators that include removal services as part of the coverage. These mechanisms have not been widely implemented but represent potential paths for financing development.

Private financing mechanisms based on the commercial value of removal services have advanced alongside the growth of commercial removal companies. Astroscale, ClearSpace, and other commercial operators have developed business models that combine government funding for demonstration missions with commercial revenue from removal services for specific customers. The commercial viability of removal services depends on customers' willingness to pay for removal, which is currently driven primarily by regulatory requirements and reputational considerations rather than by direct commercial value. The expansion of the commercial market depends on the continued development of these motivating factors.

User-fee mechanisms tied to launch activity or satellite operations have been proposed in various forms. The principle would be to collect fees from operators based on their use of orbital resources, with the proceeds funding the removal of priority targets. The political obstacles to user fees in the international context are considerable, but national-level user-fee mechanisms could serve as pilots to demonstrate the approach. U.S. discussions of various forms of orbital fees have been ongoing for years without resulting in implementation, but they have continued to evolve.

The combination of these mechanisms within a sustainable financing framework remains the central challenge. No single mechanism is likely to be sufficient, and the specific combination that emerges will depend on developments in the broader space governance ecosystem. The pace of capability development, combined with the urgency of the cascade analysis frames, creates pressure to develop effective financing arrangements over the coming decade.

Insurance Innovation and the Path Forward

The space insurance market is well-positioned to play a central role in developing more comprehensive economic frameworks for debris management. Several innovations could enhance the market's contribution to broader debris policy.

Differentiated pricing based on sustainability practices, currently in the early stages of implementation, could be expanded to provide stronger commercial incentives for debris-mitigation discipline. The expansion would depend on sufficient data to support pricing differentiation and on underwriters' willingness to apply it systematically. The Space Sustainability Rating and similar frameworks provide useful inputs, but the systematic integration of these inputs into pricing decisions has progressed slowly.

Catastrophic risk pools could provide the capacity needed for major debris-related events. Developing such pools, possibly through public-private structures that include governmental backstops for catastrophic events, could expand the financial resilience of the space industry. Various proposals have been developed, but implementation has been slow due to the complexities of cross-jurisdictional structures and limited claims experience to inform pricing.

Long-term risk products that price operators' long-term debris exposure over multi-decade horizons could provide stronger incentives for sustainable practices. Conventional insurance products primarily focus on short-term operational risks and offer limited explicit pricing of long-term debris exposure. Developing long-term risk products could fundamentally change the economic incentives for sustainability decisions, but the technical and commercial challenges are significant.

Linkage to broader environmental, social, and governance (ESG) frameworks could provide additional pathways for integrating debris considerations into commercial decisions. The growing importance of ESG considerations in institutional investment has created incentives for companies to demonstrate sustainability performance across multiple dimensions. Integrating space sustainability into ESG frameworks, through specific reporting standards and evaluation methodologies, could create additional commercial pressure to adopt debris-mitigation discipline.

The development of these innovations will require sustained engagement among the insurance and space industries, regulatory bodies, and the academic and policy communities. The pace of development has been slow but is accelerating, particularly as the debris challenge has gained broader visibility and the commercial implications have become more concrete.

The Economic Argument for Sustainability

A specific question raised by the economic analysis is whether sustainability considerations are economically aligned with commercial interests in the long run. The conventional view that sustainability requires sacrificing commercial value is increasingly being challenged by analyses suggesting that sustainable practices may yield better long-term economic outcomes than alternative approaches.

The argument has several components. First, the long-term operability of orbital regions depends on the cumulative practices of all operators in those regions. An operator who contributes to degrading a region's operability also degrades their own future operating environment and that of others. Operating in a degraded environment entails higher conjunction-avoidance costs, higher insurance premiums, higher casualty risk costs, and potentially the eventual loss of access to specific orbital regimes. These costs accrue over decades, and an operator that minimizes its contribution to degradation receives proportional benefits.

Second, the regulatory environment is tightening. Operators that have invested in sustainability practices are better positioned to comply with tightening requirements at a lower cost than those that have not. The transition from twenty-five-year disposal to five-year disposal in major jurisdictions is a specific example: operators that had already invested in shorter disposal timelines were able to comply at limited additional cost, while operators that had relied on longer timelines faced more substantial adjustments. Similar transitions in other regulatory dimensions are likely, and operators that have anticipated them are at an advantage.

Third, the commercial environment is increasingly attentive to sustainability considerations. Customers, investors, and partners increasingly evaluate space operators on sustainability metrics. Operators with strong sustainability practices are positioned to attract better customers, lower-cost capital, and stronger partnerships. Operators with weak sustainability practices face the opposite effects. The cumulative commercial impact of these metrics has been increasing and is likely to continue to do so.

Fourth, the presence of catastrophic risks, large fragmentation events, cascade scenarios, and regulatory shocks creates considerable value in maintaining the capability to operate through such events. Operators with strong sustainability practices are better positioned to maintain operations during adverse scenarios, both because their own contributions to such scenarios are smaller and because their operational practices enable them to navigate the consequences. The risk-adjusted value of sustainable practices may therefore exceed the apparent value under deterministic projections.

These arguments do not eliminate the conventional tension between immediate cost and long-term value, but they suggest that the tension is less than it might initially appear. The economic case for sustainability has been strengthening over time, and this trend is a principal driver of the broader transition toward more sustainable practices in the space industry. Whether the transition will proceed at the pace required to address the debris challenge depends on the continued strengthening of the economic case and on the development of supporting policies and institutions.

Figure 13.1. Global Space Economy by Sector, 2024–2025
(Total: approximately $570–630 billion)

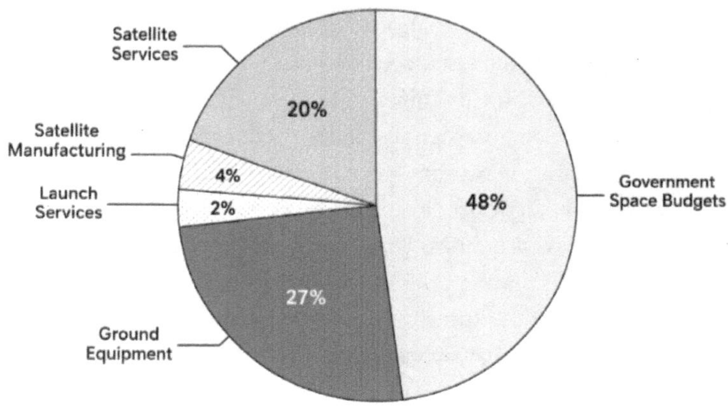

Category	Annual Premium Volume	Typical Premium Rate	Coverage Period
Launch insurance	~$200–250 million	1–3% of insured value	Launch + commissioning
In-orbit insurance	~$150–200 million	0.5–2% per year	Annual
Third-party liability	~$50–80 million	Modest; statutory minimums	Per-mission
Total market	~$400–500 million	–	–

Table 13.1. Space Insurance Market Characteristics by Coverage Category, 2024–2025

Chapter Summary

The orbital debris challenge is fundamentally an economic problem as well as a technical one, with technical solutions depending heavily on the economic frameworks within which space operators make their decisions. The global space economy operates at approximately $570-630 billion annually as of 2024-2025. It is projected to exceed $1 trillion within the next decade, providing significant financial resources for space activities while also creating strong pressure for rapid deployment, which can complicate sustainability considerations. The costs of debris mitigation, while modest as a percentage of total mission costs, aggregate to considerable absolute amounts across the industry.

The space insurance market plays a central role in pricing space-related risks, and the integration of debris considerations into pricing is advancing through differentiated pricing and supporting frameworks such as the Space Sustainability Rating program. The shared nature of orbital debris creates classic challenges in aligning individual incentives with collective interests, and mechanisms such as property rights, government regulation, user fees, and cooperative arrangements address different aspects of this structure. Financing active debris removal remains a particular challenge, as the misalignment between who pays and who benefits creates persistent obstacles to securing the scale of financing that cascade analysis suggests is required. The economic case for sustainability practices has been strengthening over time, with regulatory tightening, commercial considerations, and risk-adjusted analyses all contributing.

Whether the economic frameworks will develop at the pace required to support the technical response to the debris challenge depends on continued progress across multiple dimensions, including integrating debris considerations into insurance pricing, developing sustainable financing mechanisms for active removal, and the continued maturation of the broader space governance ecosystem.

Chapter 14. Human Spaceflight and Astronaut Safety

Introduction

The orbital debris environment poses fundamentally different consequences for crewed spacecraft than for uncrewed satellites. A debris impact that disables a communications satellite causes commercial loss and operational disruption; a debris impact that breaches the pressure hull of a crewed vehicle threatens lives. This asymmetry has shaped both the operational practices of human spaceflight programs and the technical requirements for crewed vehicles, which are far more stringent than those applied to comparable uncrewed missions.

This chapter examines the relationship between orbital debris and human spaceflight in detail. It begins with the operational history of debris-related events affecting the International Space Station and other crewed spacecraft, drawing lessons that have shaped contemporary practice. It examines the protective measures incorporated into ISS design and the operational practices that supplement them. It examines the specific debris-related considerations for commercial human spaceflight, including the Crew Dragon and Starliner programs and the developing commercial space station programs. It examines the implications of the evolving debris environment for future human spaceflight, including the longer-duration missions to the Moon and Mars contemplated by the Artemis program. It examines the broader question of how the safety of human spaceflight should be weighed within the debris policy framework.

The Debris Environment of the International Space Station

The International Space Station orbits at an altitude of approximately 408 kilometers (254 miles) in a 51.6-degree inclination orbit, completing one revolution every 90 minutes. The station has been continuously occupied since November 2000, with more than 25 years of crewed operations as of mid-2026. The orbital configuration has been chosen for both operational reasons (compatibility with the Russian launch sites at Baikonur and accessibility from Cape Canaveral and other launch facilities) and safety reasons (sufficient atmospheric drag to ensure the station decays naturally if abandoned and sufficient distance from the most heavily affected debris bands to limit collision risk).

The debris environment at 408 kilometers (254 miles) is much less dense than in the most heavily affected bands between 700 and 1000 kilometers (435-621 miles), but it is not benign. Atmospheric drag at 408 kilometers (254 miles) continuously clears debris from the local environment. However, the clearance rate depends on the size and ballistic coefficient of individual fragments. It is generally too slow to remove all fragments produced by major events at higher altitudes. Fragments from events at higher altitudes can pass through the ISS regime during their decay, and fragments from events near the station can persist locally for extended periods.

The station has performed conjunction-avoidance maneuvers at varying rates throughout its operational life. The historical rate has been about 1 to 3 maneuvers per year, with the rate increasing in the years following large debris-generating events. The 2007 Fengyun-1C anti-satellite test and the 2009 Iridium-Cosmos collision both produced fragment populations that later required ISS avoidance maneuvers. The maneuver rate has increased over time as the surrounding debris environment has grown, with the station performing several maneuvers per year in recent years.

The conjunction screening process for the ISS is more rigorous than that for typical commercial satellites because of human safety implications. Conjunctions are screened against tighter thresholds, position uncertainties are evaluated more carefully, and the decision to maneuver is made at lower probability thresholds than for uncrewed missions. NASA's Mission Control Center coordinates with Russian counterparts at the Mission Control Center-Moscow on conjunction decisions affecting the joint operation. The procedures have been refined over the station's operational life through accumulated experience and have generally been effective at preventing collisions despite the increasing debris environment.

The most direct demonstration of the operational risks posed by orbital debris to crewed spaceflight came with the November 2021 Cosmos 1408 anti-satellite test. The fragment cloud produced by the Russian test passed near the ISS orbit shortly after the intercept, prompting the seven-person crew to take shelter in their two return spacecraft, a Soyuz and a SpaceX Crew Dragon, as a precaution against potential impacts. The shelter procedure lasted about two hours and was repeated several times in the following days as the fragment cloud passed nearby on later station orbits. The episode was the most direct evidence in many years of the operational consequences of debris-generating activities, and the diplomatic response, particularly the U.S. response that followed, reflected significant concern about the precedent it set.

Whipple Shielding and Spacecraft Protection

Protection of the ISS against small-fragment impacts relies primarily on Whipple shielding and related multi-layer protective designs. The Whipple shield, named for astronomer Fred Whipple, who proposed it in 1947, consists of a thin outer plate (the bumper) separated by a stand-off distance from a thicker rear wall. When a high-velocity fragment impacts the bumper, it produces a debris cloud that disperses across the stand-off distance before reaching the rear wall. The dispersed cloud spreads the impact energy over a larger area, allowing the rear wall to absorb it without penetration.

The original Whipple shield design has been refined for ISS applications. The stuffed Whipple shield includes layers of intermediate materials (Kevlar, Nextel, and other ballistic textiles) between the bumper and the rear wall. These materials further disperse the impact debris cloud and improve the shield's protection against larger fragments. The multi-shock shield uses multiple bumpers in sequence, each producing additional dispersion of the impact debris. These advanced designs protect against fragments larger than the original Whipple shield could address.

The ISS uses a combination of these shielding designs across different modules and on external surfaces. The U.S. Destiny laboratory module incorporates extensive stuffed Whipple shielding designed to protect against fragments up to approximately 1.4 centimeters (0.55 inches) at typical ISS impact velocities. Other U.S. modules have similar protection. The Russian segments incorporate equivalent shielding designed to Russian standards. The European Columbus module and the Japanese Kibo module have shielding designs that meet ISS requirements while incorporating module-specific design choices.

The shielding is supplemented by several operational practices that further reduce risk. Critical components are typically located in shielded interior locations rather than on external surfaces. External equipment is positioned to minimize exposure to the primary directions of debris flux. Operational practices for extravehicular activities account for debris risks, and EVAs are typically scheduled during periods of reduced flux. The cumulative effect of these practices greatly reduces the risk of consequential damage from debris impacts.

The shielding's performance has been validated through both ground-based hypervelocity impact testing and in-orbit observations. Small particles have struck the ISS many times throughout its operational life, and impacts have generally caused only minor surface damage that has not affected critical functions. The largest documented impact in mid-2021 left a small hole in the Canadarm2 robotic arm. It did not affect the arm's operational capability but provided an important data point for assessing shielding performance. The impact was attributed to a small fragment whose origin could not be definitively determined.

The vulnerability of specific external components has been a recurring concern. The Quest airlock, external science instruments, solar array structures, and radiator panels have all been identified as potentially vulnerable to specific impact scenarios. These vulnerabilities have been addressed through combinations of additional local shielding, redundancy in critical functions, and operational procedures that limit exposure during high-risk periods. The continued evolution of these protective measures, in response to both accumulated operational experience and the changing debris environment, has been a recurring theme in ISS operations.

The Operational Practices of ISS Conjunction Avoidance

Operational practices for conjunction avoidance at the ISS have been progressively refined throughout the station's operational life. The current practices reflect lessons learned from accumulated experience and integration with the broader U.S. and Russian conjunction-screening infrastructure.

Conjunction warnings are typically received about three days before predicted close approaches. Both NASA and the Russian Space Agency review the warnings, focusing the assessment on the predicted miss distance, position uncertainties, and the probability of collision. The decision-making framework requires consideration of several specific factors: the probability of collision relative to operational thresholds, the consequences of a potential impact (which depend on the impact location on the station), the operational impact of a maneuver (including disruption of ongoing activities and propellant consumption), and the broader operational context.

The maneuver thresholds for the ISS are tighter than those for typical commercial satellites. NASA's general practice has been to perform avoidance maneuvers when the probability of collision exceeds approximately 1 in 100,000, with additional factors that may warrant maneuvers at lower thresholds. The Russian practice has been similar in substance, though with somewhat different specific parameters. The combined U.S.-Russian decision-making produces conjunction-avoidance practices that are generally more conservative than those applied to most commercial spacecraft.

Avoidance maneuvers require coordination across multiple operational systems. The maneuver is typically executed by firing the propulsion of one of the docked spacecraft (a Progress cargo vehicle or a Russian Soyuz vehicle) rather than the station's own propulsion, which is more limited. Maneuver planning must account for the station's mass distribution, the location and capabilities of the docked spacecraft, and the implications for subsequent operations. The maneuvers are typically small, with a delta-V of less than 3.3 ft/s (1 m/s), but require precise execution and verification.

The decision to perform an avoidance maneuver is not always made, even when conjunction warnings are received. In many cases, the predicted miss distance and the probability of collision do not warrant a maneuver, or the operational implications of maneuvering outweigh the safety benefits. Non-maneuver decisions are documented with explicit consideration of the factors involved, and the cumulative experience with these decisions informs the later development of the decision-making framework.

The shelter-in-place procedure the ISS crew followed during the November 2021 Cosmos 1408 event represents a distinct response to debris risk. When the debris environment poses an elevated risk that avoidance maneuvers cannot effectively mitigate because of the volume of fragments, the unpredictability of their motion, or the timing of events, the crew can take shelter in their return spacecraft. The procedure provides additional protection beyond the station's general shielding because the returning spacecraft are smaller targets and can be quickly evacuated to Earth if the situation deteriorates. The procedure was developed specifically for scenarios involving elevated debris risk and has been used several times over the station's operational life.

Crew Spacecraft Design Considerations

The spacecraft used to transport crew to and from the ISS faces specific debris-related design considerations that differ from those for either uncrewed satellites or the ISS itself. The vehicles spend relatively brief periods in orbit, typically days to weeks for crew rotation missions, but the consequences of debris impact during those periods are particularly severe.

The SpaceX Crew Dragon, in operational service since 2020, incorporates multiple protective measures against debris impacts. The pressure vessel is designed with structural margins that exceed operational requirements, protecting it against minor impacts that could otherwise damage less robustly designed vehicles. The thermal protection system, while designed primarily for reentry heating, also provides incidental protection against small-fragment impacts during orbital operations. The trunk section, jettisoned before reentry, provides additional structural mass to absorb impacts without affecting the crew compartment.

The Boeing Starliner, which conducted its first crewed mission to the ISS in 2024, incorporates similar protective measures. The vehicle has faced operational challenges during its development and initial operational missions. However, the basic protective approach to debris impact has remained consistent with the standards established for U.S. crew vehicles. The Russian Soyuz, which has been the primary Russian crew vehicle since the 1960s, has been progressively upgraded over its operational life and incorporates protective measures consistent with Russian standards for crewed spacecraft.

The Chinese Shenzhou spacecraft, used for missions to the Tiangong space station, incorporates protective measures developed within the Chinese space program. The specific design details have not been the subject of detailed international examination, but the overall pattern is consistent with the protective approaches used by other crew vehicle programs.

Operational practices for crew vehicles in transit between the ground and the space station include specific provisions for debris-related events. Launch trajectories are coordinated with conjunction screening to identify potential conflicts with cataloged objects. On-orbit operations between launch and station docking are similarly screened. Reentry trajectories are planned to avoid known debris populations and minimize time spent in orbital regimes with elevated debris risk.

The development of new commercial crew vehicles for applications such as space tourism, commercial space stations, and lunar missions has continued in recent years. Each new vehicle has incorporated debris-related design considerations consistent with the broader requirements of crewed spacecraft, though the specific approaches have varied. The continued evolution of these design practices, integrated with the developing operational practices of commercial human spaceflight, will be a key area of debris policy development over the coming decade.

Commercial Space Stations and Debris Considerations

The development of commercial space stations marks a new chapter in human spaceflight, with specific implications for debris policy. The International Space Station is nearing the end of its operational life, with retirement and controlled deorbit scheduled for around 2030. NASA's strategy to replace the ISS's operational capacity includes developing multiple commercial space stations through the Commercial Low Earth Orbit Destinations program.

Several commercial space station concepts are under development. Axiom Space has been building a commercial space station that will initially attach to the ISS and later operate as a free-flying station after the ISS is retired. The first Axiom module is planned for launch in the late 2020s. Sierra Space, in partnership with Blue Origin, has been developing the Orbital Reef commercial station concept. Voyager Space has been developing the Starlab commercial station concept. Each program is at a different stage of development, with operational deployment expected in the late 2020s and early 2030s.

Debris-related considerations for commercial space stations are similar in substance to those applied to the ISS but require specific implementation in each program's design. Protective measures must address the station's specific orbital environment, operational profile, and crew accommodations. Conjunction screening processes must be tailored to each station's specific operations and coordinated with both U.S. government tracking services and the operators of other space stations and commercial satellites in the same orbital regions.

The selection of orbital environments for commercial stations has implications beyond debris considerations. The ISS orbit, at an altitude of 408 km (254 miles) and a 51.6° inclination, has provided a workable balance of operational and safety considerations for more than two decades. Commercial stations may select different orbits based on their specific operational requirements, with corresponding implications for the debris environment in which they operate. Stations operating at higher altitudes would face a higher debris risk relative to the ISS; stations operating at lower altitudes would face greater atmospheric drag and corresponding propulsion requirements. The trade-offs will be specific to each program.

The integration of commercial stations into the broader regulatory framework for debris management has been a major focus of U.S. regulatory development. The FAA, FCC, and other agencies have been developing frameworks for commercial human spaceflight, including provisions for operating commercial stations. These frameworks reflect the broader trend of expanding debris-related regulation into new areas of space activity.

End-of-life considerations for commercial stations are especially important from a debris policy perspective. Each commercial station will eventually reach the end of its operational life and require disposal, and that disposal must be managed to minimize debris-related consequences. Controlled reentry, similar to the planned ISS deorbit, is the standard approach for large stations. Developing operational capabilities for the controlled deorbiting of commercial stations, including the required propulsion systems and operational practices, has been a key aspect of commercial station development.

The ISS Deorbit

The end of operations for the International Space Station, planned for around 2030, will be one of the most historic debris-related events in space history. The station has a mass of approximately 420 metric tons (463 tons), larger than any object previously deliberately disposed of. The disposal must be managed to ensure that the station reenters in a manner that minimizes both the risk of casualties on Earth and the consequences of any residual debris.

The plan for ISS deorbit, developed jointly by NASA and Roscosmos, calls for a controlled reentry using a dedicated U.S. Deorbit Vehicle (USDV). In 2024, NASA contracted with SpaceX to develop the USDV, a modified Crew Dragon vehicle with substantial additional propulsion. The USDV will dock with the station near the end of its operational life and, after the final crew departure, perform a series of deorbit burns that lower the station's perigee until atmospheric drag accelerates the descent into a controlled reentry.

The deorbit sequence for a 420-metric-ton object is unlike anything previously attempted. The U.S. Deorbit Vehicle will need to provide approximately 100 to 150 meters per second of total delta-V to lower the station's perigee from its operational altitude of roughly 408 kilometers to an entry interface altitude of approximately 120 kilometers. Given the station's mass, this requires a vehicle with a wet mass of several tens of metric tons of propellant, operating with high reliability during a period when the station itself will be unpowered and passivated. The sequence will be executed over several days, with each burn progressively lowering the orbit until atmospheric drag takes over and completes the descent. The target impact corridor in the South Pacific Ocean Uninhabited Area is expected to be approximately 8,000 kilometers long and 1,500 kilometers wide, centered well south of any inhabited territory.

The selected impact zone is the South Pacific Ocean Uninhabited Area (SPOUA), the same general region used for the disposal of the Mir station in 2001 and for many subsequent satellite disposals. Its remote location, far from any inhabited landmasses, offers the best available balance between minimizing casualty risk and operational feasibility. The reentry footprint is expected to span several thousand kilometers (miles) along the descent trajectory, with the largest surviving fragments expected to impact in the central portion of the SPOUA.

The technical challenges of the ISS deorbit are substantial. The station's mass is sufficient that the deorbit propulsion requirements exceed those of any previous operational deorbit. The station's structural complexity, with multiple pressurized modules, large external structures, and various external equipment, poses challenges to maintaining a controlled attitude throughout the deorbit sequence. The interactions between the USDV and the station during deorbit must be precisely managed to prevent damage that could disrupt controlled reentry.

The debris consequences of the ISS deorbit have been the subject of serious analysis. The station will produce surviving fragments that reach Earth's surface, with the largest expected to weigh on the order of metric tonnes. The total surviving mass is expected to be on the order of 50-100 metric tonnes (55-110 tons), distributed across hundreds of fragments. The atmospheric impact of the reentry will be significant, with the station's structural materials, primarily aluminum, contributing to the metallic aerosol loading of the upper atmosphere discussed in Chapter 6.

The transition from ISS operations to the operational deployment of commercial stations marks a major period of change in human spaceflight. The cumulative debris implications of the transition, including the deorbit of the ISS, the deployment of commercial station modules, the operations of crew vehicles supporting both, and the broader expansion of human spaceflight, will be significant and require careful management. The development of operational practices and regulatory frameworks specifically for this transition has been an active area in the years leading up to the ISS retirement.

Implications for Lunar and Cislunar Human Spaceflight

The Artemis program and other lunar exploration efforts contemplate human spaceflight beyond low Earth orbit. The debris-related considerations for these missions differ in important respects from those for LEO operations and represent a developing aspect of debris policy.

The cislunar environment, examined in more detail in Chapter 16, currently has a much smaller debris population than low Earth orbit. The number of objects in cislunar space is on the order of dozens, compared with tens of thousands in LEO. The collision risk for cislunar operations is correspondingly lower in absolute terms. However, the consequences of any debris generation in cislunar space are significant because debris persists for very long times.

Protecting crewed lunar missions from debris is therefore primarily a matter of careful trajectory planning and operational practices rather than the heavy shielding required for LEO operations. The Orion spacecraft, designed for the Artemis missions, incorporates protective measures suited to the cislunar environment. The Gateway lunar space station, being developed for cislunar operations, will be the most prominent new crewed asset in this orbital environment.

Longer-duration missions associated with lunar exploration, and eventually with potential Mars missions, introduce additional debris-related considerations. The crew habitats, transit vehicles, and surface assets will be exposed to debris environments that differ dramatically from those in LEO. Protective measures must be tailored to these specific environments, with corresponding implications for vehicle design and operational practice.

Debris-related practices for lunar and cislunar human spaceflight have evolved alongside the broader Artemis architecture. NASA, ESA, JAXA, and other partners have been developing the specific design and operational practices required. Integration with the broader international framework for cislunar activity, including the Artemis Accords discussed in Chapter 7, provides the policy context for these practices.

The eventual development of human Mars missions, planned for the 2030s or later, will face additional debris-related challenges. The transit between Earth and Mars passes through diverse debris environments, including the LEO debris belt on departure and the much sparser interplanetary environment during transit. Arrival at Mars will encounter the small but growing population of debris in Mars orbit, which has accumulated from robotic missions over several decades. The development of debris-related practices for human Mars spaceflight is in its early stages, but it will be a key aspect of debris policy as missions approach.

The Safety Argument in Debris Policy

The role of human safety considerations in debris policy has been a recurring topic of discussion. The safety implications of debris for crewed spacecraft are unambiguous; debris impacts pose lethal risks that must be addressed through specific protective measures. The question is how to weigh these safety implications against other considerations within the broader debris policy framework.

One position holds that human safety should be paramount, with debris policy designed primarily to ensure the continued safety of human spaceflight. The argument is that debris-related casualties in human spaceflight would have severe consequences for both the affected individuals and the broader human spaceflight enterprise, and that the policy framework should therefore prioritize preventing such casualties. This position has strong support in the human spaceflight community and has shaped specific aspects of operational practice.

A second position holds that human safety considerations should be one of many factors within the broader debris policy framework. The argument is that debris-related risks affect all space activities, including predominantly uncrewed satellite operations, which constitute the majority of space activity. The policy framework should therefore address debris risks comprehensively, with human spaceflight considerations integrated into the broader framework rather than dominating it. This position has strong support within the broader space policy community and has shaped many of the developing policy frameworks discussed in earlier chapters.

In practice, reconciling these positions has required a combination of approaches. The human spaceflight community has developed specific operational practices and protective measures to address the elevated risks of crewed operations. The broader debris policy framework has incorporated human spaceflight considerations as an important factor in policy design, without focusing exclusively on them. Together, these approaches have produced practices that protect human spaceflight while also addressing the broader debris challenge.

The continued growth of commercial human spaceflight, including space tourism, commercial station operations, and eventual operations beyond LEO, will greatly expand the human spaceflight footprint in the coming decades. The increased exposure of humans to the debris environment, combined with the growing volume of crewed spacecraft, will likely create additional pressure for protective measures and for a broader debris policy that addresses human safety. The development of frameworks that effectively integrate human safety considerations into the broader debris policy will be a key aspect of policy development over the coming decade.

Lessons from ISS Operations

The accumulated experience from more than 25 years of continuous human presence on the ISS has yielded countless lessons that inform current debris policy—several of which merit discussion.

Protective measures for crewed spacecraft can be effective even in environments with substantial debris. The ISS has not experienced consequential debris impacts despite operating in LEO for 25 years and despite the ever-growing amount of debris during that period. The shielding designs, conjunction-avoidance practices, and the operational discipline the program has maintained have collectively provided effective protection. The lesson is that human spaceflight in challenging debris environments is feasible with appropriate measures, but those measures must be specifically designed and operationally maintained.

The integration of multiple protective approaches is essential. No single measure, whether shielding, conjunction avoidance, operational discipline, or shelter-in-place, provides comprehensive protection against the diverse debris-related risks. The combined application of multiple measures, each addressing specific aspects of the risk profile, produces more effective protection than any single measure could provide. The lesson is that comprehensive protection requires a system of integrated measures rather than reliance on any single approach.

Cooperation among partners has been essential to ISS operations. U.S.-Russian cooperation, the foundation of ISS operations, has continued despite the deterioration of broader bilateral relations after 2014 and the further deterioration after 2022. Operational practices for debris management, including conjunction screening and avoidance maneuver decision-making, have continued to function effectively despite the political deterioration. The lesson is that operational cooperation in space can be maintained even amid political tensions, but it requires sustained institutional infrastructure and operational practice.

Operational transparency has been valuable for both safety and policy purposes. The ISS program has been more transparent than many other space programs in documenting operational events, including conjunction maneuvers, shelter-in-place events, and impact incidents. Transparency has supported both immediate safety considerations and a broader policy understanding of debris-related risks. The lesson is that operational transparency, while sometimes uncomfortable for the operators involved, contributes to the broader management of space-related risks.

The continued evolution of practices in response to changing conditions is essential. The ISS protective measures and operational practices have been progressively refined throughout the station's operational life, in response to both accumulated experience and the evolving debris environment. The static application of practices established at the program's beginning would not have been sufficient to maintain safety amid changing conditions. The lesson is that debris-related practices must be continuously developed and refined, with explicit institutional capacity to incorporate new understanding into operational practice.

The Future of Human Spaceflight in a Crowded Sky

The combination of growing debris populations, expanding human spaceflight, and the institutional and technical responses to both poses specific challenges for the coming decade. Several developments will shape the trajectory.

The deployment of commercial space stations, beginning in the late 2020s and continuing through the 2030s, will expand the human presence in LEO. Cumulative crew time on commercial stations is expected to exceed historical ISS time within the next decade, and this expansion will require the development of protective measures and operational practices.

The expansion of space tourism, particularly through suborbital operations and short-duration orbital missions, will introduce new categories of human spaceflight participants. These participants are typically less prepared than professional astronauts to handle debris-related risks of space activity, and operational practices must accommodate this difference. The development of frameworks specifically addressing space tourism debris remains in its early stages.

The Artemis program and other lunar exploration efforts will establish a sustained human presence in cislunar space, driving the development of debris-related practices for that region. The transition from intermittent to sustained cislunar human spaceflight, expected to occur progressively over the coming decade, will require comprehensive new development in debris-related practices.

The development of human missions to Mars, while still far in the future, will introduce additional dimensions to debris-related practice. The Mars debris environment is currently small but growing, and eventual human missions will require careful consideration of how to manage debris-related risks across the diverse environments encountered during those missions.

The cumulative effect of these developments is that human spaceflight will expand considerably in scope and duration during a period when the debris environment is also expanding rapidly. Integrating human spaceflight safety with the broader debris policy framework will require ongoing development to ensure both objectives can be achieved together. Lessons from the ISS experience, combined with the specific considerations of new types of human spaceflight, provide the foundation for this integration. Whether the integration proceeds at the pace required to support the expanding human spaceflight enterprise depends on continued attention from both the human spaceflight community and the broader space policy community.

Figure 14.1. International Space Station Stuffed Whipple Shielding Configuration
(U.S. segment cross-section, schematic; not to scale)

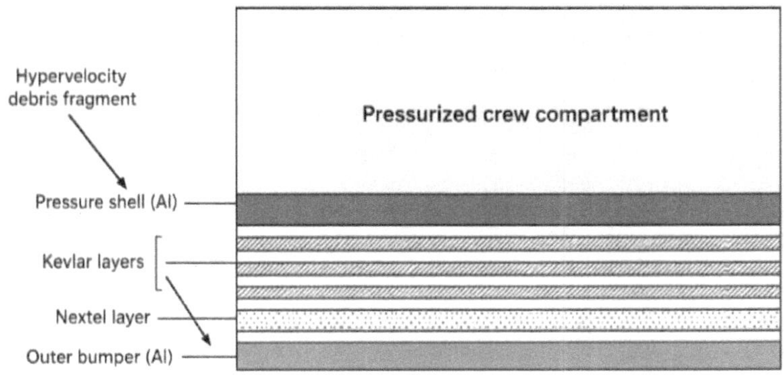

Year	Maneuvers	Notable Events
2008	1	Pre-Iridium-Cosmos baseline
2009	0	Iridium-Cosmos collision (no ISS impact)
2014	1	Routine
2020	3	Increased rate; Long March debris
2021	3	Cosmos 1408; crew sheltered
2022	2	Long March 6A debris
2023	4	Multiple constellation conjunctions
2024	4	Resurs-P1 debris; ongoing operations
2025	3	Continued elevated rate

Table 14.1. ISS Conjunction Avoidance Maneuver History (Selected Years)

Chapter Summary

The orbital debris environment poses fundamentally different risks to crewed spacecraft than to uncrewed satellites, with impacts that threaten lives rather than merely disrupt operations. The International Space Station, in operation since November 2000 at an altitude of approximately 408 km (254 miles), has demonstrated that human spaceflight in challenging debris environments is feasible with appropriate protective measures. The station's protection relies on Whipple shielding and related multi-layer designs, supplemented by operational practices such as conjunction avoidance and shelter-in-place procedures.

The 2021 Cosmos 1408 anti-satellite test provided the most direct demonstration in many years of the operational risks debris poses to crewed spaceflight, prompting the seven-person ISS crew to take shelter in their return spacecraft as the fragment cloud passed nearby. The crew vehicles used for ISS operations, including Crew Dragon, Starliner, Soyuz, and Shenzhou, incorporate protective measures consistent with the requirements for crewed spacecraft. The development of commercial space stations under NASA's Commercial Low Earth Orbit Destinations program, with operational deployment expected in the late 2020s and early 2030s, will greatly expand human presence in LEO. The planned deorbit of the ISS around 2030, to be executed by a dedicated U.S. Deorbit Vehicle being developed by SpaceX, will be one of the most significant debris-related events in space history.

The expansion of human spaceflight beyond LEO through the Artemis program and eventual Mars missions introduces additional debris-related considerations that are being progressively addressed. Lessons from ISS operations, including the effectiveness of integrated protective measures, the importance of cooperation among partners, the value of operational transparency, and the necessity of the continued evolution of practices, provide the foundation for the ongoing development of debris-related practices in human spaceflight. Integrating human spaceflight safety with the broader debris policy framework will be one of the developing challenges of the coming decade as the orbital environment grows more crowded and human spaceflight expands in scope and duration.

Chapter 15. Astronomy, Environment, and the Orbital Commons

Introduction

The orbital debris challenge has been treated throughout most of this book as a technical problem involving objects in orbit and the consequences of their interactions. But the cumulative environmental impact of space activity extends beyond the orbital regime itself. The visibility of operational satellites and debris fragments to ground-based astronomy has emerged as a substantial concern over the past several years, particularly with the deployment of large constellations whose members are individually visible to the naked eye and collectively interfere with both visible-light and radio-frequency observations. The atmospheric impact of satellite reentries, examined in Chapter 6, represents another dimension of cumulative environmental impact. The nighttime sky itself, considered a shared cultural and scientific resource, has been altered in ways with implications that extend well beyond the technical concerns of orbital operators.

This chapter examines the broader environmental dimensions of space activity. It begins with the impact of satellite constellations on ground-based astronomy, including both optical and radio-frequency effects. It examines the specific concerns of the astronomical community and the responses developed by satellite operators. It examines the broader question of preserving the nighttime sky as a cultural and scientific issue. It examines the developing concept of the orbital commons and the implications for governance frameworks that must address the cumulative environmental impact of space activity. The treatment is integrative, drawing on the technical material from earlier chapters while extending the analysis to the broader environmental and cultural dimensions of the debris challenge.

The Impact of Satellite Constellations on Optical Astronomy

The deployment of large satellite constellations beginning in the late 2010s has increased the number of bright objects visible in the nighttime sky. Each Starlink satellite has an apparent magnitude of about 3 to 5 under typical observing conditions, making them readily visible to the naked eye and significantly brighter than most natural celestial objects. The OneWeb satellites, operating at higher altitudes, are typically less bright but still readily detectable by astronomical equipment. The Amazon Leo, Guowang, and Qianfan constellations are deploying satellites with similar visibility characteristics.

Professional astronomy's working threshold for problematic satellite contamination is approximately magnitude 7, the limit of naked-eye visibility under dark-sky conditions. Objects brighter than magnitude 7 appear in virtually any ground-based exposure of sufficient duration. Starlink satellites in their initial deployments, before orbital raising, were frequently between magnitude 2 and 4, comparable to the brightest stars in the sky and dramatically brighter than the magnitude 7 threshold. Even after SpaceX's VisorSat dimming measures brought some satellites down to magnitudes of 5.5 to 6.5, they remained above the working threshold for many telescope programs. The Vera C. Rubin Observatory's technical team concluded that a single 30-second exposure of the sky during twilight hours will, on average, capture one or more satellite trails at the brightness levels of the current constellation environment, with that rate rising to five to ten trails per exposure as deployments scale.

The cumulative effect on the appearance of the nighttime sky is readily apparent. At dark-sky locations, satellites are now visible in large numbers throughout the night. During twilight, when satellites at higher altitudes are illuminated by the sun while observers on the ground are in darkness, many hundreds of satellites can be seen simultaneously. The visual impression of the sky has been altered in ways that affect both casual observation and professional astronomy.

The impact on professional astronomy varies by observation type. Wide-field surveys, which photograph large portions of the sky to identify and track many objects simultaneously, are particularly affected because individual exposures frequently capture satellite tracks. The Vera C. Rubin Observatory in Chile, which began full operations in 2025 with the Legacy Survey of Space and Time, has been extensively analyzed for the expected impact of satellites. Estimates suggest that approximately 30-50 percent of long-exposure images taken during twilight hours will contain visible satellite tracks, significantly degrading the quality of the resulting science data.

Specific scientific programs have been particularly affected. Detecting near-Earth asteroids requires wide-field surveys at twilight, when both the asteroid and the observatory are illuminated by sunlight, exactly the conditions under which satellite contamination is highest. Detecting transient phenomena, such as supernovae, gamma-ray bursts, and gravitational-wave electromagnetic counterparts, requires rapid follow-up observations that satellite passes may disrupt. Characterizing solar system objects, which often relies on observations at specific times and orientations, can be significantly affected by satellite interference.

The mitigation of optical impact has been pursued through several approaches. Satellite operators have developed darker satellite designs that reduce the reflected sunlight that makes satellites visible. SpaceX's DarkSat and VisorSat designs, deployed beginning in 2020, incorporated specific measures to reduce satellite brightness. Subsequent generations of Starlink satellites have continued to incorporate brightness-reduction measures. The reductions have been notable, typically half a magnitude or more, but have not eliminated the visibility of the satellites.

Operational practices that minimize impact have also been developed. Orienting solar panels during specific orbital phases can reduce reflected light during the most visibility-sensitive periods. Adjusting the timing of satellite operations can minimize exposure during astronomical observation periods. Coordinating orbital deployments to avoid the most disruptive configurations has also been pursued. These practices have produced incremental reductions in impact but have not eliminated the notable impact of large constellations on optical astronomy.

The astronomical community has been increasingly organizing to address the impact on the constellation. The International Astronomical Union established the Center for the Protection of the Dark and Quiet Sky from Satellite Constellation Interference in 2022, providing institutional infrastructure for technical analysis and policy advocacy. The Center has produced detailed analyses of constellation impacts and has engaged satellite operators, regulators, and policymakers on mitigation approaches. The American Astronomical Society and other national astronomical organizations have similarly developed policy positions and engagement strategies.

Radio Astronomy and Spectrum Allocations

The impact of satellites on radio astronomy is, in some respects, more severe than their optical impact because radio observations are often more sensitive to interference, and the affected frequencies cannot be readily filtered. Modern satellite constellations transmit at frequencies that overlap with portions of the spectrum used for astronomical observation, and their cumulative interference can effectively eliminate radio observations in those bands.

The Square Kilometer Array, the most ambitious radio astronomy facility under construction with sites in South Africa and Australia, has been particularly concerned about the constellation's impact. The facility's observation bands include frequencies that overlap with constellation downlinks, and its interference rejection design has been influenced by the expected constellation environment. The facility's full operations, expected to progress through the late 2020s and early 2030s, will face a markedly altered operating environment due to constellation deployments.

Radio astronomy protection has been addressed primarily through the international Radio Regulations administered by the ITU, which allocate specific frequency bands to radio astronomy and limit interference from other services. This protection has been broadly effective for the historical bands allocated to radio astronomy. However, the rapid deployment of constellations has created new categories of interference that the historical framework did not fully anticipate. The 2023 World Radiocommunication Conference adopted several decisions to protect radio astronomy from interference from constellations, but implementation has been piecemeal, and their practical effectiveness remains under evaluation.

The Atacama Large Millimeter Array (ALMA) in Chile and the Very Large Array (VLA) in New Mexico are among the major radio astronomy facilities affected by constellations. Each facility has developed specific approaches to managing interference, including time-based scheduling to avoid the most disruptive constellation passes and data processing techniques that can remove certain types of interference. The cumulative impact remains severe, and the long-term trajectory of radio astronomy in a constellation-dominated environment is uncertain.

The relationship between satellite operators and the radio astronomy community has encompassed both cooperation and tension. Several operators have engaged constructively with the astronomical community on mitigation strategies, including specific operational practices that reduce interference during sensitive observation periods. Other operators have been less engaged, with corresponding consequences for affected facilities. The development of broader cooperative frameworks, analogous to those that have emerged for optical impact, has progressed but remains incomplete.

The Cultural Dimension of the Nighttime Sky

Beyond the specific impacts on professional astronomy, the alteration of the nighttime sky has cultural dimensions that affect a much broader population. The visible nighttime sky has been a fundamental aspect of human experience throughout history, with cultural, religious, scientific, and aesthetic significance across virtually all societies. The transformation of the sky by artificial light pollution from ground sources has been ongoing for more than a century, significantly affecting the cultural relationship between humans and the night sky. The new contributions from satellite constellations represent an additional transformation occurring much more rapidly than historical changes.

The dark-sky movement, organized for several decades to address ground-based light pollution, has begun to extend its focus to satellite-related impacts. The International Dark-Sky Association, the leading global organization in this field, has incorporated constellation issues into its broader advocacy. The movement's perspective frames the nighttime sky as a shared resource analogous to clean air or water, with corresponding implications for the appropriate management of activities that affect it.

The position enjoys considerable support in cultural and environmental policy circles, though it has not generally been incorporated into specific regulatory frameworks. The argument is that the visible nighttime sky has intrinsic value that should be considered when managing activities that affect it, regardless of whether those activities have specific economic or scientific justifications. The position parallels other environmental positions that emphasize intrinsic values rather than instrumental ones, and it has produced specific policy proposals, including absolute limits on satellite brightness, no-deployment zones in areas of cultural significance, and protections for indigenous and other cultural relationships with the sky.

The implementation of these positions within regulatory frameworks has been limited. Most major jurisdictions have not adopted specific regulations addressing the cultural impact of satellite deployments. The U.S. Federal Aviation Administration's environmental review processes for commercial launches have begun to address cumulative environmental impacts. Still, the inclusion of cultural sky impact has been limited. ESA's mission planning processes incorporate some consideration of cumulative impact, but the cultural dimension has been less prominent than other considerations.

The continued development of regulatory frameworks that incorporate cultural sky considerations is a key aspect of debris and space policy. Pressure for such development comes from cultural and environmental advocacy organizations, from affected indigenous communities and others with specific cultural relationships to the sky, and from elements of the scientific community who link concerns about astronomical impacts to broader cultural considerations. The pace and direction of development remain uncertain, but the cultural dimension is likely to receive greater attention in the coming years.

The Atmospheric Impact Revisited

The atmospheric impact of satellite reentries, examined in detail in Chapter 6, represents another dimension of cumulative environmental impact that extends beyond the orbital environment. The 2023 NOAA measurements, which show that approximately 10 percent of stratospheric aerosol particles contain spacecraft-derived metals (Murphy et al., 2023), have established an empirical foundation for substantial concern about the cumulative atmospheric impact of space activity.

The trajectory of atmospheric impact, given the projected growth in satellite reentries, is among the most consequential aspects of the broader environmental dimension of space activity. As discussed in Chapter 6, the steady-state mass of satellite material reentering Earth's atmosphere each year is expected to grow exponentially over the coming decade as large constellations reach operational scale and their members are progressively replaced through normal turnover. The projected annual reentry mass of several thousand tonnes will far exceed the historical baseline and may approach or even exceed the natural meteoritic input to the atmosphere.

The implications for atmospheric chemistry, as discussed in Chapter 6, are still being assessed. The interaction of metallic aerosols with stratospheric chemistry, the implications for ozone, contributions to climate forcing, and other specific impacts are all subjects of active research. The empirical foundation provided by the NOAA measurements has prompted extensive new research, which is being progressively published, and the picture continues to develop.

The relationship between atmospheric impact and broader environmental considerations is important. Atmospheric impact is closely linked to satellite end-of-life disposal practices, with implications for both surface-level casualty risk and atmospheric chemistry. The integration of these considerations into the broader policy framework has progressed but remains incomplete. The development of integrated environmental analysis that addresses both ground and atmospheric impacts is an emerging aspect of debris policy.

Engagement between the atmospheric science community and debris policy has been growing. Research programs at NOAA, NASA, the European Space Agency, and various academic institutions are addressing the atmospheric implications of space activity. Integrating these findings into the broader policy framework requires sustained engagement between the atmospheric science and space policy communities. While this engagement has progressed, it remains in its early stages.

The Space Sustainability Movement

Concerns about debris, astronomical and atmospheric impacts, and other environmental dimensions of space activity have fueled a growing space sustainability movement. The movement encompasses multiple organizations, technical programs, policy initiatives, and academic research efforts that share a common framing of space activity as having significant environmental dimensions that require explicit management.

The Space Sustainability Rating program discussed in Chapter 13 is one specific manifestation of the movement. The Center for the Protection of the Dark and Quiet Sky, mentioned earlier in this chapter, is another. The various international initiatives in COPUOS and other forums on long-term sustainability are part of the movement's broader institutional infrastructure. Academic research programs that integrate technical analysis with policy considerations, advocacy organizations that engage with regulators and operators, and affected communities that articulate cultural and environmental concerns all contribute to the movement.

The movement's development has been substantial over the past decade. The 2019 adoption of the Long-term Sustainability Guidelines by COPUOS marked a significant institutional milestone. The 2022 establishment of the Center for the Protection of the Dark and Quiet Sky marked another milestone. The progressive incorporation of sustainability considerations into national regulatory frameworks, examined in Chapter 8, represents the practical implementation of the broader framing. The continued development of the movement, both as an organizational network and as an intellectual framework, has been a principal pattern in debris and space policy over the past decade.

The challenges facing the movement are formidable. The cumulative environmental impact of space activity spans multiple dimensions: orbital debris, astronomical, atmospheric, and cultural impacts. Different communities with distinct concerns and institutional infrastructures address these dimensions. The integration of these dimensions into a coherent sustainability framework has progressed but remains incomplete. The political and economic obstacles to substantive policy development that meaningfully addresses the cumulative impact are formidable, and progress has been slower than the underlying growth of space activity.

The relationship between the space sustainability movement and broader environmental movements has been a notable feature of its development. Space sustainability concerns have increasingly been integrated with those related to climate change, biodiversity loss, and other environmental issues, as the broader environmental community recognizes space sustainability as an important dimension of cumulative environmental impact. This integration has provided both broader institutional support for space sustainability concerns and connections to the more developed institutional infrastructure of the broader environmental movement.

The Concept of the Orbital Commons

The cumulative environmental considerations discussed in this chapter have produced a developing conceptualization of orbital space as a common, shared resource that no individual operator owns, from which no individual operator can be excluded, and whose degradation from individual use accumulates to affect all users. The concept has been progressively developed in academic and policy literature and has begun to influence specific policy frameworks.

The orbital commons concept parallels other commons concepts in environmental governance, including the atmospheric commons (addressed through climate policy), the oceanic commons (addressed through the Law of the Sea and various specific regimes), and the polar commons (addressed through the Antarctic Treaty system). Each of these commons has produced specific governance frameworks that address the unique characteristics of the resource and the activities that affect it. The orbital commons concept implies that similar frameworks should be developed for orbital space, with corresponding institutional and substantive elements.

The application of commons reasoning to orbital space has produced several specific policy proposals. Property rights frameworks, in which specific use rights are assigned to operators with corresponding responsibilities, have been proposed but face significant implementation challenges. Use fees, with proceeds funding the preservation of the commons, have been discussed in various forms. International governance frameworks with the institutional capacity to address commons-level concerns have been proposed but not yet substantively developed. Each of these approaches addresses specific aspects of the broader commons challenge.

The most direct application of commons reasoning has been in the developing argument for international governance of long-term sustainability. The argument is that the cumulative environmental impact of space activity, including but not limited to debris, has the structure of a commons problem that no individual state's regulatory framework can fully address. Therefore, developing international governance with effective authority over cumulative impacts is necessary for effective long-term management of the orbital environment.

The implementation of comprehensive international governance has been the most difficult aspect of the broader response to common concerns. The treaty-level instruments developed in earlier decades, examined in Chapter 7, do not provide the binding authority that effective commons governance requires. The non-binding instruments, including the LTS Guidelines and the various technical guidelines, lack the binding character necessary for effective enforcement. The growing recognition of the common character of orbital space has not yet produced governance arrangements adequate to address it.

The continued development of the orbital commons concept, both as an intellectual framework and as a basis for specific policy development, will be a principal trajectory of debris policy over the coming decade. The integration of the various dimensions of cumulative environmental impact into a coherent commons framework, the development of governance arrangements that can substantively address commons concerns, and the connection of these developments to broader environmental governance will collectively shape that trajectory.

Indigenous Perspectives and the Sky

A specific dimension of the cultural impact of space activity that deserves explicit attention is the relationship between indigenous communities and the nighttime sky. Many indigenous cultures around the world have specific relationships with the sky that are integral to their cultural identity, religious practice, and traditional knowledge. The alteration of the visible sky by satellite constellations has implications for these relationships, which indigenous communities and their advocates have progressively articulated.

The relationship is particularly strong in regions where modern light pollution from terrestrial sources has been less severe, and the natural sky has remained relatively intact. The communities in these regions have maintained traditional astronomical knowledge, ceremonial practices involving specific celestial alignments, and cultural narratives that depend on the sky's unaltered appearance. The introduction of satellite contamination affects all of these dimensions.

The Mauna Kea telescope debates in Hawaii, while focused on ground-based astronomical infrastructure rather than satellite impacts, have surfaced important issues regarding the relationship between Western astronomical activities and indigenous cultural practices. Native Hawaiian opposition to the Thirty Meter Telescope was rooted in part in concerns about the mountain's cultural significance and the broader question of who has authority to make decisions about activities affecting culturally significant resources. Comparable issues also arise in the context of satellite impacts, where similar questions regarding authority, responsibility, and consent emerge under different operational circumstances.

Engagement between satellite operators, regulators, and indigenous communities has been limited. Decision-making for constellation deployments has generally not included specific consultation with indigenous communities, even those whose cultural relationships with the sky may be deeply affected. Developing consultation processes, perhaps modeled on practices that have emerged in terrestrial projects affecting indigenous communities, is an increasingly important aspect of space governance.

The legal frameworks for protecting indigenous interests in space-related activities are limited. International instruments such as the United Nations Declaration on the Rights of Indigenous Peoples (UNDRIP) provide some framework, but their application to space activities has not been comprehensively developed. National frameworks for indigenous rights vary widely and have generally not been applied to space activities. The development of legal frameworks that address indigenous interests in space-related activities is an emerging aspect of broader space governance.

The deeper question is how the cultural dimensions of space activity, including but not limited to indigenous concerns, should be integrated into the broader governance framework. The conventional framing of space governance has emphasized technical, economic, and security considerations, leaving cultural dimensions largely overlooked. The progressive development of frameworks that incorporate cultural considerations, including indigenous perspectives, will be a significant development in space governance over the coming decade.

The Astronomical Community's Engagement

The astronomical community's response to the impact of constellations has evolved rapidly over the past several years. This evolution provides a model for broader engagement of affected communities in space governance. Several specific developments merit attention.

The development of technical analysis capabilities within the astronomical community has been extensive. Detailed analyses of the impact of specific constellations on specific observation programs, modeling of cumulative impact across the projected constellation environment, and assessment of mitigation effectiveness have all advanced. These analyses provide the empirical foundation for both immediate operational decisions and broader policy advocacy.

The institutional infrastructure for astronomical engagement with space governance has been developing. The IAU Center for the Protection of the Dark and Quiet Sky, established in 2022, serves as a focal point for the international astronomical community's engagement. The American Astronomical Society's policy committee, the European

Astronomical Society's equivalent body, and similar national organizations provide additional institutional infrastructure. Coordination among these organizations, supported by their parent astronomical societies, has produced increasingly coherent engagement with the policy community.

The sustained engagement with satellite operators has yielded mixed results. Some operators, particularly SpaceX, have been developing mitigation measures to address optical impacts. Other operators have been less engaged. The variation in engagement reflects both operator characteristics and the specific issues at stake. The astronomical community's continued engagement, with progressive refinement of technical analysis and advocacy approaches, has produced incremental improvements but has not eliminated the underlying impact.

Engagement with regulatory bodies has progressed but remains in its early stages. The astronomical community has provided input to the FCC, the ESA, the ITU, and other regulatory bodies regarding the impact of constellation deployments. The incorporation of astronomical considerations into specific regulatory decisions has been inconsistent, with some decisions explicitly considering astronomical impacts and others largely ignoring them. The development of regulatory frameworks that systematically address astronomical impacts remains in its early stages.

International coordination of astronomical engagement has been particularly important. The impact of the constellation affects astronomy globally, and the response requires international coordination across the astronomical community. The IAU and other international astronomical organizations have provided the institutional infrastructure for this coordination. The continued development of coordinated approaches will be important for the broader effectiveness of the community's engagement.

Integration with Broader Environmental Governance

The cumulative environmental impact of space activity across debris, astronomy, atmosphere, and cultural dimensions suggests the value of integrating space environmental considerations into broader environmental governance frameworks. Several specific connections merit attention.

The connection to climate governance is significant. The atmospheric impact of space activity, including both launch emissions and reentry effects, may have implications for climate forcing. The integration of these considerations into broader climate governance, through frameworks such as the Intergovernmental Panel on Climate Change and the United Nations Framework Convention on Climate Change, has been limited but is progressing. The development of integrated approaches that address both terrestrial and space-related contributions to climate considerations is an emerging aspect of environmental governance.

The connection to biodiversity governance is less direct but still exists. The impact of light pollution, both terrestrial and space-related, on wildlife, particularly migratory birds and nocturnal species, has been documented and represents a specific dimension of cumulative environmental impact. Integration with broader biodiversity governance, through frameworks such as the Convention on Biological Diversity, has been limited but offers another potential avenue.

The connection to sustainable development governance is significant. The 2030 Agenda for Sustainable Development, with its specific Sustainable Development Goals, provides a framework that encompasses environmental, social, and economic dimensions. The integration of space sustainability with the broader Sustainable Development Goals framework is evolving, with specific connections to goals related to climate (Goal 13), terrestrial ecosystems (Goal 15), and broader sustainability concerns (Goal 17 on partnerships).

The connection to environmental impact assessment frameworks is potentially important. Many jurisdictions have developed comprehensive frameworks for environmental impact assessment across various activity categories, and the application of these frameworks to space activities is progressing. The U.S. National Environmental Policy Act has been applied to specific commercial space activities, with implications for cumulative impact considerations. ESA's environmental impact frameworks similarly address aspects of space activities. The progressive integration of space activities into broader environmental impact assessment frameworks could advance the systematic consideration of cumulative impact.

The institutional integration of space environmental considerations into broader environmental governance has progressed slowly. The space governance institutions, including COPUOS, the ITU, and IADC, are largely separate from the broader environmental governance institutions. Developing more integrated institutional frameworks, perhaps through enhanced coordination among existing institutions or through new bodies spanning the relevant domains, is a developing possibility for the coming decade.

Toward Comprehensive Sustainability

Integrating the dimensions discussed in this chapter, including orbital debris, astronomical impact, atmospheric impact, cultural impact, and broader environmental considerations, into a comprehensive sustainability framework is one of the principal challenges of contemporary space policy. Developing such a framework would require substantial progress in several specific areas.

Developing integrated technical analysis capabilities that address all dimensions of cumulative impact is a key challenge. The current technical analysis is fragmented across specialized communities, each developing capabilities for its specific concerns. Integrating these capabilities with consistent methodologies and shared empirical foundations would greatly strengthen the empirical basis for sustainability policy.

Developing institutional infrastructure capable of addressing comprehensive sustainability is another challenge. The current institutional landscape addresses different aspects of sustainability through separate bodies with distinct mandates and operational modes. Integrating these arrangements, perhaps through enhanced coordination among existing bodies or through new institutional development, would improve the practical capacity for comprehensive sustainability policy.

Developing policy frameworks that incorporate comprehensive sustainability considerations is a substantive challenge. Current frameworks address sustainability through specific instruments that focus on particular dimensions. Developing more comprehensive frameworks, including treaty instruments, national regulations, and operational practices that address the multiple dimensions of cumulative impact, would considerably improve the policy basis for sustainability.

The development of effective implementation mechanisms is the final challenge. The various policy frameworks that may emerge will require effective implementation through national regulatory frameworks, operators' operational practices, and economic mechanisms that align individual incentives with collective sustainability. Developing these implementation mechanisms is an ongoing challenge that depends on continued progress across multiple dimensions.

The pace of development across these dimensions has varied. Technical analysis has progressed reasonably well across most dimensions. Institutional development has been slower. Policy framework development has advanced in some areas but more slowly in others. Implementation mechanisms have developed primarily through national-level frameworks, with limited international integration. Continued progress, integrated across the various dimensions, will determine whether the broader sustainability framework can advance at the pace required to address the considerable environmental challenges posed by space activity.

Figure 15.1. Projected Impact of Large Constellations on the Vera C. Rubin Observatory: Observation Programs by Time of Night

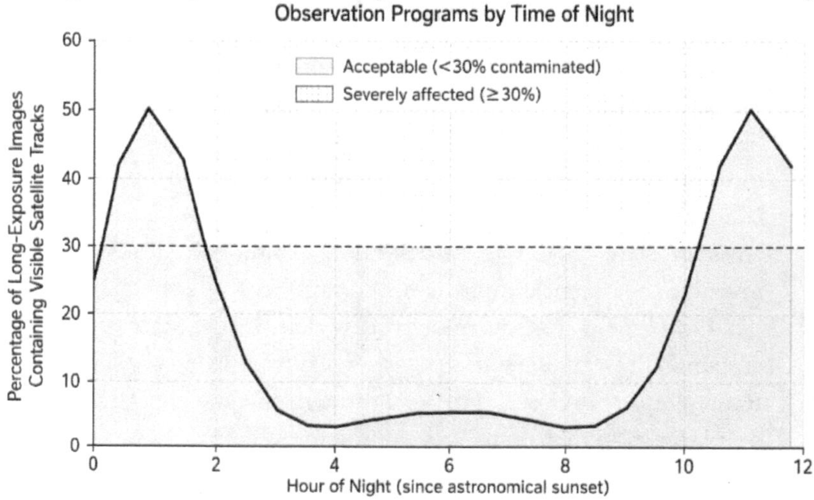

Facility	Type	Location	Principal Concern
Vera C. Rubin Observatory	Wide-field optical survey	Cerro Pachón, Chile	Twilight track contamination
Square Kilometer Array (SKA)	Radio array	South Africa, Australia	Downlink interference
Atacama Large Millimeter Array	Radio interferometer	Atacama, Chile	mm-wave interference
Very Large Array (VLA)	Radio interferometer	New Mexico, USA	Downlink interference
Hubble Space Telescope	Space optical	Earth orbit	Track contamination at low altitudes
Subaru / Keck / Gemini	Ground optical	Mauna Kea, HI	Track contamination

Table 15.1. Major Astronomical Facilities Affected by Constellation Deployments

Chapter Summary

The cumulative environmental impact of space activity extends well beyond the orbital regime, with far-reaching implications for ground-based astronomy, the atmosphere, and the cultural and aesthetic relationship between humans and the nighttime sky. The deployment of large satellite constellations has dramatically increased the number of bright objects visible in the nighttime sky, with a pronounced impact on both optical and radio astronomy. The Vera C. Rubin Observatory and the Square Kilometer Array, among other major astronomical facilities, face considerable challenges from the constellation environment, which have been progressively analyzed and addressed through various mitigation approaches. The atmospheric impact of satellite reentries, established empirically by the 2023 NOAA measurements showing measurable accumulations of spacecraft-derived metals in stratospheric aerosols, represents another dimension of cumulative impact that is projected to grow steadily as the reentry rate from large constellations increases.

The cultural impact on indigenous communities and others with specific relationships to the nighttime sky represents another important dimension that has been increasingly articulated. The space sustainability movement has been developing as an organizational network and intellectual framework that integrates these concerns, supported by institutional developments such as the IAU Center for the Protection of the Dark and Quiet Sky and the Space Sustainability Rating program.

The concept of orbital space as a common, analogous to other environmental commons, has been progressively developed and provides a framework for more comprehensive governance arrangements. The integration of space environmental considerations with broader environmental governance frameworks remains in its early stages but represents an important direction for future development. Developing comprehensive sustainability frameworks that address all dimensions of cumulative impact is one of the principal challenges of contemporary space policy, with progress required across technical analysis, institutional infrastructure, policy frameworks, and implementation mechanisms.

Chapter 16. Cislunar Futures and the Path Forward

Introduction

The orbital debris challenge examined in this book extends well beyond the Earth-orbiting regime that has been its principal focus. The progressive expansion of human and robotic activity beyond low Earth orbit, through the Artemis lunar exploration program, the development of cislunar infrastructure, and eventual missions to Mars and other destinations, introduces new dimensions to the debris challenge that existing frameworks have only begun to address. The cislunar environment, currently containing only dozens of objects, will see rapid growth in the coming years, and lessons learned from managing the Earth-orbiting environment must be applied to prevent the cislunar orbit from following the same trajectory toward congestion and risk.

This final chapter examines the evolving cislunar environment and the path forward for debris management more broadly. It begins by outlining the technical characteristics of cislunar orbit and the debris-related considerations that distinguish it from the Earth-orbiting environment. It then examines emerging patterns of cislunar activity, including the Artemis program, commercial lunar missions, and potential future activity. It reviews early efforts to develop a cislunar debris policy, including the Artemis Accords provisions and emerging technical standards. It considers how lessons from Earth-orbiting debris management should inform the development of cislunar policy. And it concludes with a synthesis of the book's principal themes and an assessment of the path forward for orbital debris management.

The Cislunar Environment

The cislunar orbit encompasses the volume of space between Earth and the Moon, including the Earth-Moon Lagrange points, the stable and unstable orbits around those points, lunar polar and low orbits, and the trajectories that connect Earth orbits to lunar space. The dynamics of the cislunar environment are considerably more complex than those of Earth-orbiting regimes because the gravitational influences of both Earth and the Moon must be accounted for, with the Sun introducing additional perturbations.

The technical characteristics of cislunar space differ from those of Earth-orbiting regimes in several important respects. The relevant timescales are longer, and orbital periods range from days to weeks rather than minutes to hours. The relevant velocities also differ, with lunar altitudes considerably lower than those in low Earth orbit. Atmospheric drag, which clears low Earth orbit debris within years to decades, does not operate in cislunar space, where the persistence times of any debris generated would extend essentially indefinitely. The radiation environment is harsher, with far less protection from Earth's magnetosphere.

The current population in cislunar space is small. Approximately 60-80 objects are in some form of cislunar orbit, including operational lunar missions, decommissioned probes, and debris from past missions. The objects include the Lunar Reconnaissance Orbiter, China's Queqiao relay satellite at the Earth-Moon L2 point, several CAPSTONE-class small satellites in the Near Rectilinear Halo Orbit being established for Artemis operations, the Korean Pathfinder Lunar Orbiter, India's Chandrayaan-3 vehicle elements, and other operational and defunct objects.

The historical accumulation of debris in cislunar space has been very small compared with the Earth-orbiting environment. The principal sources have been spent upper stages from lunar missions, occasional collisions or fragmentation events involving lunar probes, and ejecta from lunar surface impacts caused by probes deliberately impacted at the end of their missions. The cumulative effect has been to populate cislunar space with a modest number of large objects, whose orbits have been characterized by various tracking efforts.

Tracking cislunar objects is more challenging than tracking Earth-orbiting objects. The greater distances and lower angular rates of cislunar objects produce distinct observational characteristics, and the optical and radar systems used for Earth-orbit tracking are not well optimized for cislunar applications. Specialized cislunar tracking capabilities have been developed through several programs, including the U.S. Cislunar Highway Patrol System concept and various academic and commercial initiatives. Developing a comprehensive cislunar tracking infrastructure is a massive undertaking that will progress over the coming decade.

The Artemis Program and Lunar Activity

The Artemis program, NASA's plan for sustained human and robotic exploration of the Moon, is the most consequential ongoing development affecting the cislunar environment. The program is structured around a series of progressively capable missions. Artemis I, an uncrewed test flight of the Space Launch System and Orion spacecraft, was successfully conducted in late 2022. Artemis II launched on April 1, 2026, carrying NASA astronauts Reid Wiseman, Victor Glover, and Christina Koch, along with Canadian Space Agency astronaut Jeremy Hansen, on a ten-day journey around the Moon and back, splashing down in the Pacific Ocean on April 10, 2026. In 2026, NASA announced that Artemis III, rather than attempting a lunar landing, would conduct an Earth-orbit rendezvous-and-docking demonstration between Orion and the program's privately developed lunar landers, with a target date of 2027. The plan shifts the first crewed Moon landing to the Artemis IV mission, targeting 2028, with NASA Administrator Jared Isaacman stating the agency is also pursuing a possible second landing under Artemis V in the same year. Both SpaceX, with its Starship Human Landing System, and Blue Origin, with its Blue Origin lander, are under contract to support the landing missions. Subsequent missions in the program will progressively expand operations to include sustained surface activities and the construction of the Lunar Gateway space station in the Near Rectilinear Halo Orbit around the Moon.

The Lunar Gateway is a key component of the Artemis architecture and a critical new feature of the cislunar environment. The Gateway will be a small space station orbiting the Moon in a Near Rectilinear Halo

Orbit, providing infrastructure for crewed and robotic operations on and around the Moon. The first Gateway elements, the Power and Propulsion Element and the Habitation and Logistics Outpost, are planned for launch in 2027 and will be progressively assembled through subsequent module deployments. The Gateway is being developed through international cooperation among NASA, ESA, JAXA, and the Canadian Space Agency, with each agency contributing specific elements.

International participation in the Artemis program has been extensive. The Artemis Accords, opened for signature in October 2020, have been signed by more than 50 states as of mid-2026. The accords establish principles for cooperative lunar exploration, including specific provisions on space debris and orbital sustainability. The participation of multiple states in the program creates both opportunities for cooperation and governance complexities, particularly as the program scales up to sustained lunar surface operations.

Beyond the U.S.-led Artemis program, several other major lunar exploration efforts are underway. The Chinese International Lunar Research Station, in cooperation with Russia and several other partners, has advanced through preparatory missions and is planned for operational deployment in the late 2020s and early 2030s. The Russian Luna program has continued with several recent missions, including the 2023 Luna-25 mission, which failed to land. The Indian Chandrayaan program has continued through Chandrayaan-3 in 2023, with subsequent missions in development. The Japanese SLIM lander conducted its lunar landing in early 2024. The South Korean lunar program has progressed through preparatory activities. Each of these national programs contributes to the broader cislunar activity and the corresponding development of debris-related considerations.

The commercial lunar sector has been developing rapidly. Several U.S. companies, including Astrobotic, Intuitive Machines, and Firefly Aerospace, have conducted commercial lunar missions or have missions in advanced development. The NASA Commercial Lunar Payload Services program has been a critical contractor for these missions, supporting the development of commercial capabilities while providing payload delivery services to NASA's science programs. The Intuitive Machines Odysseus mission in 2024 was the first commercial lunar surface landing. Follow-on commercial missions have continued through 2024 and 2025, with multiple additional missions in development.

Cislunar Debris Considerations

The debris-related considerations for cislunar activity differ fundamentally from those for Earth-orbiting activity, and the existing frameworks must be adapted to the cislunar environment. Several specific considerations warrant attention.

The persistence of debris in cislunar space is essentially indefinite on policy-relevant timescales. Atmospheric drag, which clears low Earth orbit debris within years to decades, does not operate in cislunar space. Once a debris fragment is created in cislunar space, it remains there until it impacts the Moon, is captured by Earth and reenters the atmosphere, escapes the Earth-Moon system, or is removed through active intervention. The natural pathways are limited and slow, so the practical implication is that the cislunar environment is far more sensitive to debris generation than the Earth-orbiting environment.

The dynamics of cislunar debris are more complex than those of Earth-orbiting debris. The Earth-Moon Lagrange points and the stable and unstable orbits around them produce distinct patterns of object motion that depend on multiple gravitational influences. Debris fragments generated in cislunar space can drift through different dynamic regimes over time, making their eventual trajectories difficult to predict precisely without detailed simulation. The implications for tracking, conjunction assessment, and active management are considerable.

The collision risk in current cislunar operations is low because the population is small. Still, the implications of any collision are severe due to the persistence of debris and the difficulty of its removal. A collision generating thousands of fragments at low Earth orbit altitude would degrade that altitude band, but the fragments would clear over decades to centuries; an equivalent collision in cislunar space would produce fragments that would persist effectively forever and could affect operations across multiple cislunar orbital regions. The asymmetry between the consequences in the two regimes is pronounced.

Mitigation practices developed for Earth-orbiting operations need to be adapted for cislunar applications. End-of-life disposal options for cislunar missions differ. Lunar surface impact has been a common practice for spent stages and decommissioned probes, but it can produce debris that may affect future surface operations. Solar orbit injection has been used as an alternative disposal approach for some missions and avoids contaminating the cislunar regime, but it requires large amounts of propellant. Heliocentric orbits between Earth and Mars or beyond may be appropriate for some missions, but they pose specific tracking and management challenges.

Conjunction screening for cislunar operations is in its early stages. The U.S. Space Force has been developing cislunar tracking capabilities through several programs, and the Cislunar Highway Patrol System concept has advanced through preliminary studies. Integrating cislunar tracking into the broader space situational awareness infrastructure is an emerging challenge. Commercial providers, including LeoLabs and others, have begun expanding their capabilities into the cislunar regime.

Lunar Surface Considerations

The lunar surface presents debris-related considerations distinct from those of the orbital environment. Debris from past lunar missions, including descent stages, impact ejecta, equipment left on the surface, and other items, has accumulated over several decades. The Apollo program left much of its equipment on the surface, including descent stages, lunar rovers, and scientific instruments. Later robotic missions have added to the surface debris. The cumulative debris on the lunar surface includes hundreds of metric tons of material distributed across various landing sites.

The implications of lunar surface debris differ from those of orbital debris in several respects. The debris does not pose a collision risk to other surface operations except in the immediate vicinity of the affected sites. The debris does not degrade the broader lunar environment in ways that affect access to the Moon. The debris may have specific implications for the preservation of historic sites, the management of activities at planned operational sites, and the broader question of how lunar surface activities should be coordinated.

The Artemis Accords address surface considerations through several specific provisions. The accords commit signatories to preserving outer space heritage, including human-made artifacts on celestial bodies. They also commit signatories to deconflicting their activities, including specific provisions for establishing "safety zones" around lunar surface activities.

The implementation of these provisions is in its early stages and will be developed progressively as lunar surface activity expands.

Longer-term considerations for lunar surface activity include managing surface debris from sustained operations, assessing the implications of resource extraction for the broader lunar environment, and developing frameworks to coordinate activities among multiple actors at shared locations. The expansion of lunar surface activity through Artemis, the Chinese International Lunar Research Station, and various commercial programs will increase the volume and complexity of these considerations over the coming decade.

Cislunar Governance Frameworks

The development of governance frameworks specifically addressing cislunar activity has progressed alongside operational development. Several developments warrant attention.

The Artemis Accords, as discussed above, provide a framework that includes specific provisions for cislunar activity. The accords' provisions on debris, deconfliction, and sustainability are more detailed than those in older space treaties and reflect the specific characteristics of cislunar operations. Implementation of the accords through national-level operational practices and broader coordination among signatory states is progressing.

Although developed primarily for Earth-orbiting operations, the COPUOS LTS Guidelines include provisions that apply to cislunar activity. The committee has been considering developing additional guidelines specifically addressing cislunar operations, with discussions ongoing as of 2026. The eventual adoption of cislunar-specific guidelines would advance the international framework.

The IADC has been considering extending its mitigation guidelines to cislunar operations. The current guidelines were developed primarily for Earth-orbiting applications, but the underlying technical principles translate readily to cislunar contexts with appropriate adaptations. The committee's work to extend the framework has been advancing through its specialized working groups.

National regulatory frameworks for cislunar activity are under development. The U.S. has been working through various agencies to develop frameworks for commercial cislunar activity, including specific provisions for the Artemis program and commercial lunar missions. ESA has been developing frameworks for European cislunar activity. Other major spacefaring states have been developing their national approaches in similar ways. The variation in national frameworks for cislunar activity is potentially wide-ranging, and the international coordination of these frameworks will be an evolving aspect of cislunar policy.

The legal questions that arise in cislunar activity, including the legal status of resources extracted from the Moon, the legal framework for activities at common locations, and the legal arrangements for emergencies and assistance, are complex and have been addressed progressively through both the Artemis Accords and various national frameworks. The Moon Agreement of 1979, despite its limited ratification, provides a historical reference point, but the substantive frameworks governing cislunar activity are being developed primarily through the Accords and national legislation.

The Path Forward for Earth-Orbit Debris Management

The development of cislunar policy is one specific dimension of the broader path forward for debris management. The Earth-orbiting environment, the principal focus of this book, faces ongoing challenges that will require sustained attention over the coming decade and beyond. The synthesis of the book's principal themes provides a framework for assessing the path forward.

The technical foundation has been developed over the past several decades. Cascade analysis provides the analytical framework for understanding the long-term evolution of the debris environment. The empirical foundation, supported by the global tracking infrastructure, provides the data on which all aspects of policy depend. Technical capabilities for mitigation, tracking, conjunction screening, and active removal have advanced significantly. Continued development of these capabilities is essential, but the basic technical foundation is in place.

The policy framework has evolved significantly but remains incomplete in important respects. The international guidelines provide the technical baseline, the national regulatory frameworks provide the substantive implementation, and the various coordination mechanisms provide the operational infrastructure. The trend toward more stringent national requirements has been positive, but variation across jurisdictions remains considerable. Developing more harmonized international approaches, supported by enhanced institutional infrastructure, remains a challenge.

The economic structure has been progressively addressed through the development of insurance markets, sustainability ratings, and various financing mechanisms. The orbital environment's fundamental commons character continues to pose challenges in aligning individual incentives with collective interests. Developing more comprehensive economic frameworks that address the cumulative environmental impact of space activity remains an important challenge.

The institutional infrastructure has been developing through various international and national bodies, industry organizations, and academic and research communities. The integration of these elements into a coherent governance framework remains incomplete. Developing more integrated institutional arrangements with improved coordination across the various bodies and dimensions is one of the principal challenges for the coming decade.

Beyond the immediate orbital debris concerns, the cumulative environmental impact has emerged as a major new dimension. The astronomical impact of constellations, the atmospheric impact of reentries, and broader cultural and environmental considerations all extend the debris challenge into a more comprehensive sustainability framework. Developing policies and practices that address the full scope of cumulative environmental impact will require continued progress across multiple dimensions.

The Generational Challenge

The orbital debris challenge is, in important respects, a generational one. The decisions made by the current generation of operators, regulators, and policymakers will shape the orbital environment for decades and centuries to come. The fragments produced by today's activities will, in many cases, persist longer than the institutions that produced them. The lessons learned from managing the current environment will inform the development of frameworks for future activities that today's policies are creating space for.

The historical perspective is sobering. The fragments from the 2007 Fengyun-1C anti-satellite test will affect the orbital environment for centuries, long after the political circumstances that produced the test have faded into history. The fragments from the 2009 Iridium-Cosmos collision continue to require avoidance maneuvers by the current generation of operators, who had no role in the events that produced them. The accumulation of legacy debris from earlier eras of operations creates persistent challenges for current operations and will continue to do so for future generations.

The current generation faces specific responsibilities that historical perspective makes clear. Decisions about constellation deployment, anti-satellite testing, mitigation requirements, and active debris removal will have consequences extending for decades and even centuries. The frameworks established now will shape the activities of future operators and affect future generations' access to orbital space. The choices about how to balance immediate operational and economic interests against long-term sustainability will determine whether future generations inherit a usable orbital environment or one that has been progressively degraded.

The question of intergenerational responsibility has increasingly been articulated within the broader discussion of debris policy. The classical arguments for intergenerational ethics in environmental governance, namely that the current generation should not unduly degrade resources that future generations will need, apply directly to the orbital environment. The implementation of these arguments through specific policy mechanisms has progressed but remains incomplete.

The institutional implications of the generational challenge are far-reaching. The institutions that govern space activity have generally been organized around shorter timescales than the consequences of space activities require. The treaty instruments developed during the Cold War have generally functioned, but their adaptation to the dramatically changed circumstances of contemporary space activity has been limited. Developing institutions with an appropriate temporal perspective, capable of making decisions whose consequences will be felt over centuries rather than years, is one of the emerging challenges of debris policy.

The Emerging Picture of Space Sustainability

The synthesis of the various dimensions discussed throughout this book points toward an emerging picture of space sustainability that has evolved considerably from earlier conceptions. Several specific characteristics of this emerging picture deserve articulation.

Sustainability is integrative rather than fragmented. The earlier treatment of orbital debris as a specialized technical concern, separate from broader environmental and policy considerations, has been progressively replaced by an integrative perspective that recognizes connections among orbital debris, atmospheric and astronomical impacts, and broader environmental and cultural dimensions. This integrative perspective yields both a better understanding of the cumulative impact of space activity and more effective policy responses.

Sustainability requires international cooperation, though cooperation will be imperfect. The earlier hopes for comprehensive international treaty arrangements have been progressively tempered by the practical experience of multilateral negotiation in the current geopolitical environment. The current approach emphasizes developing guidelines, norms, and best practices to be implemented through national-level frameworks, with international coordination that supports but does not require full multilateral consensus on every issue. The approach has been largely effective for many issues, but its limits have been illustrated by major incidents in recent years, including the 2021 Cosmos 1408 test.

Sustainability requires the engagement of multiple actors beyond the traditional space community. The earlier view of space activity as the domain of national space agencies and a small number of commercial operators has been replaced by recognition that its consequences affect a much broader range of interests. The astronomical and atmospheric science communities, indigenous and cultural advocacy organizations, and the broader environmental movement all have enormous stakes in space sustainability. Developing effective frameworks requires engaging all these communities, with corresponding implications for the institutional infrastructure of space governance.

Sustainability requires both ambitious goals and pragmatic implementation. The technical analysis suggests that significant action is needed to stabilize the orbital environment in the most heavily affected altitude bands. Political and economic constraints on action are also considerable. Reconciling these requires both a clear articulation of the goals and pragmatic development of implementation mechanisms that can make progress despite the constraints. The policy work of the past decade has been heavily focused on this reconciliation, with implementation progress often achieved through limited, incremental approaches that nonetheless contribute to the broader trajectory.

Sustainability requires ongoing learning and adaptation. The orbital environment is changing rapidly, driven by the deployment of large constellations, the emergence of new types of operations, and the accumulation of impacts that are creating conditions not fully accounted for in existing frameworks. Continued development of frameworks that can adapt to changing conditions, integrate new understanding, and respond to new challenges is essential for sustained progress.

The Decisions That Lie Ahead

Several specific decisions in the coming years will shape the trajectory of debris policy and the broader orbital environment. The choices among these decisions have implications that extend across the dimensions discussed throughout this book.

The continued development or erosion of the international norm against destructive anti-satellite testing is one such decision point. The 2022 U.S. unilateral commitment to refrain from destructive direct-ascent anti-satellite testing, subsequently endorsed by more than thirty other states through United Nations resolutions, marks an important development. Whether the norm holds, whether it is extended to additional states, and whether it is effectively maintained amid ongoing strategic competition among leading powers will strongly shape the trajectory of the debris environment. The norm's effectiveness depends on continued political support and operational restraint, neither of which can be taken for granted.

The pace of development of active debris removal capability is another decision point. The current trajectory of demonstration missions, including ClearSpace-1, PRELUDE, and similar programs, provides the technical foundation, but transitioning to sustained operational capability requires considerable additional investment. Decisions by national space agencies, commercial operators, and international bodies on how to support this transition will determine whether removal capability meets the technical requirements identified by the cascade analysis.

The implementation of tighter mitigation requirements across the major spacefaring states is a third decision point. The trend toward five-year disposal requirements has been positive, but jurisdictional variation remains extensive, and compliance with these requirements depends on sustained operational discipline. Continued development of national frameworks, supported by international coordination, will determine the rate at which new debris generation is constrained.

The integration of debris considerations into broader environmental governance is a fourth decision point. Progressive integration with climate, biodiversity, and sustainability frameworks has begun, but the deeper integration that effective management of cumulative impacts would require has not yet developed. The decisions about how space sustainability is positioned within the broader environmental governance ecosystem will materially affect the institutional and intellectual support for debris policy.

The development of cislunar policy is a fifth decision point. The Artemis Accords and emerging technical standards provide the initial framework, but the substantive development of cislunar debris policy remains in its early stages. Decisions on how to address the unique characteristics of the cislunar environment, the persistence of any debris generated, the complexity of the dynamics, and international participation in cislunar activities will determine whether the cislunar environment is managed more sustainably than the Earth-orbiting environment has been.

The financial resilience of major constellation operators constitutes a sixth decision point that existing policy frameworks have yet to address clearly. A constellation of several thousand satellites managed by a solvent, well-resourced company poses a substantially different debris risk profile than the same constellation abandoned in orbit after the operator enters bankruptcy or ceases operations. The historical record offers a cautionary example of these risks. When the original Iridium LLC filed for bankruptcy in 1999, its 66-satellite constellation remained operational only because a buyer was secured quickly. Without that acquisition, 66 large satellites in polar orbit at approximately 780 kilometers could have become uncontrolled debris, with no responsible operator available to manage their disposal. Current regulatory frameworks require disposal planning before launch, but generally do not require escrow accounts, bonding, or other financial safeguards capable of funding disposal independently of an operator's continued solvency. As satellite constellations expand into the thousands, the financial scale of potential disposal liabilities increases correspondingly, highlighting a regulatory gap that policymakers have only recently begun to address seriously.

Conclusion: The Final Frontier and Its Stewardship

The orbital environment is the first global commons that human technology has both created and threatened. The achievement of this environment, through six decades of expanding space activity, has been one of the great accomplishments of human civilization. The threats to it, through the cumulative consequences of that same activity, are now large and visible. The choices made by the current generation will determine whether the environment is preserved for future generations or progressively degraded by inadequate stewardship.

The technical analysis examined throughout this book clarifies the situation. The debris environment in the most heavily affected altitude bands is at or near the cascade threshold. The continued growth of activity, particularly through large commercial constellations, places additional pressure on an already-stressed environment. The mitigation requirements that have been progressively developed are necessary but not sufficient to stabilize the environment. The active removal capability that the cascade analysis suggests is required has progressed through demonstration, but has not yet reached the operational scale needed.

The policy and institutional analysis is similarly clear. The international frameworks established during the Cold War era have provided a foundation, but are inadequate to address contemporary challenges. National regulatory frameworks have been tightening progressively, but remain inconsistent across jurisdictions. Economic structures continue to externalize the costs of debris-generating activities. The institutional infrastructure for international coordination is inadequate to meet the demands of contemporary space activity.

The path forward requires continued development across all of these dimensions. The technical capabilities must continue to mature, with particular emphasis on active debris removal at an operational scale. The policy frameworks must continue to develop, with particular emphasis on harmonization across national frameworks and on integrating debris considerations into broader environmental governance. The institutional arrangements must continue to evolve, with particular emphasis on building capacity for international coordination on shared challenges. The economic structures must continue to develop, with particular emphasis on financing mechanisms for sustainability investments and on aligning commercial incentives with collective interests.

The challenge is formidable, but the response is increasingly capable as well. The growing community of operators, regulators, researchers, and advocates engaged with debris policy has produced progress that would have been difficult to achieve in earlier decades. International institutions have been progressively addressing debris-related questions despite broader strategic tensions. The commercial sector has been progressively investing in sustainability practices, driven by both regulatory requirements and recognition of long-term commercial interest. The astronomical, atmospheric science, and broader environmental communities have been progressively engaging with debris policy. Each of these elements contributes to the broader response.

Whether the response will be adequate to address the challenge is the central question this book has examined. The honest answer, as of mid-2026, is that the question remains open. The progress of the past decade has been exceptional, but the trajectory of orbital activity has accelerated even more. The development of the response in the coming decade will determine whether the orbital environment is stabilized or whether cascade dynamics produce the degradation that the technical analysis identifies as possible.

The orbital environment is, in the language often used to describe it, the final frontier of contemporary technological civilization. The frontier offers extraordinary opportunities for communication, scientific discovery, economic development, and human exploration. It also imposes responsibilities for environmental stewardship, for preserving access for future generations, and for managing the cumulative consequences of human activity. Reconciling opportunity and responsibility is the fundamental challenge of orbital debris management. This balance has been evolving through the work examined in this book. Whether it will continue to evolve at the pace and scale the challenge requires is the question the coming decade will answer.

The stewardship of the orbital environment is, ultimately, a test of contemporary civilization's capacity to manage the consequences of its technological achievements. That capacity has been demonstrated in some respects, such as the development of mitigation practices, the establishment of international frameworks, and the engagement of multiple communities in addressing the challenge. It has been inadequate in other respects, including the persistent commons character of the environment, gaps in international governance, and limited progress on active removal at scale. The development of that capacity in the coming years will determine whether the orbital environment becomes a success story of contemporary environmental governance or a failure.

The choice rests with the current generation. The decisions taken now, on anti-satellite testing, mitigation requirements, investment in active removal, international cooperation, and the integration of space activity with broader environmental governance, will shape the orbital environment for decades and centuries. These decisions are being made in real time by individual operators, regulators, policymakers, and citizens, whose collective choices constitute the response to the challenge. An understanding of the challenge, supported by the analysis presented in this book and the extensive body of related work, provides the foundation for informed decision-making.

The orbital environment will reflect the quality of our stewardship. Fragments from today's activities will testify to today's choices for centuries. The frameworks established now will determine what future generations can achieve. Integrating orbital debris management with broader environmental sustainability will set precedents for addressing future technological frontiers. The work continues, and the outcomes remain uncertain.

The future of space depends on the decisions we make today.

—Jason P. Crutchfield

Figure 16.1. Earth–Moon cislunar regime showing the five Lagrange points and a representative Near Rectilinear Halo Orbit (NRHO) used for the Lunar Gateway concept.

(Schematic illustration; distances and orbital dimensions not to scale.)

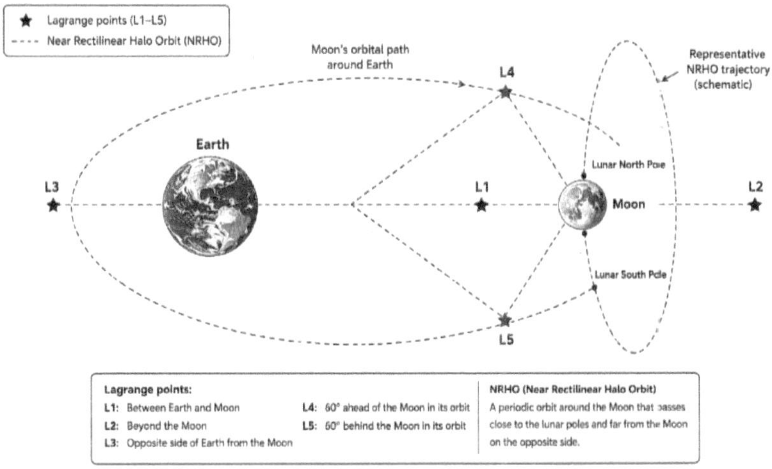

Mission	Operator	Launch / Status	Purpose
CAPSTONE	NASA / Advanced Space	2022	NRHO Pathfinder for Gateway
Artemis I	NASA	2022 (completed)	Uncrewed lunar test flight
Chandrayaan-3	ISRO	2023	Lunar south pole landing
Luna-25	Roscosmos	2023 (failed)	Lunar lander
SLIM	JAXA	2024	Precision lunar landing
IM-1 (Odysseus)	Intuitive Machines	2024	Commercial lunar landing (CLPS)
Artemis II	NASA	2026 (planned)	Crewed lunar flyby
Artemis III	NASA	2026–2028 (planned)	Earth-orbit rendezvous-and-docking demonstration

Mission	Operator	Launch / Status	Purpose
Artemis IV	NASA	2028 (planned)	Crewed lunar landing
Lunar Gateway PPE/HALO	NASA / ESA / partners	2027 (planned)	NRHO space station
China ILRS	CNSA / Roscosmos / partners	2028+ (planned)	International Lunar Research Station

Table 16.1. Active and Planned Cislunar Missions, 2020–2030

Chapter Summary

The cislunar environment, currently containing only a few dozen objects, is both a new frontier for space activity and a new challenge for debris management. The technical characteristics of the regime, including the essentially indefinite persistence of any debris generated, complex multi-body dynamics, and limited atmospheric clearing, make it more sensitive to debris generation than the Earth-orbiting environment. The Artemis program and other lunar exploration efforts will expand cislunar activity over the coming decade, with the Lunar Gateway providing critical infrastructure beginning around 2027, followed by progressive expansion toward sustained surface operations. International participation through the Artemis Accords and bilateral arrangements provides the foundation for cislunar governance, although further development is needed to address the regime's specific characteristics. Lessons from Earth-orbit debris management remain relevant to the cislunar environment but require adaptation to its specific characteristics.

The path forward for orbital debris management requires continued development across technical capabilities, policy frameworks, institutional infrastructure, economic structures, and the integration of cumulative environmental considerations. The challenge is generational, with decisions made now shaping the orbital environment for decades and centuries to come. The response has advanced considerably over the past decade, but whether it will be adequate to the challenge remains an open question. The stewardship of the orbital environment is a test of contemporary civilization's capacity to manage the consequences of its own technological achievements, and the development of that capacity in the coming years will determine whether the orbital environment becomes a success story of environmental governance or a failure. The choices rest with the current generation, and the work continues.

List of Figures

List of Tables

Appendix A. Timeline of Major Events in Space Debris History

1957 - October 4: Sputnik 1 launched, beginning the satellite era.

1968 - The Soviet Istrebitel Sputnikov co-orbital anti-satellite program begins testing and continues through 1982.

1969 – July 20: Apollo lands the first humans on the Moon, leaving the first long-term human-made objects on the lunar surface.

1972 - The Convention on International Liability for Damage Caused by Space Objects (Liability Convention) is adopted.

1976 - The Convention on Registration of Objects Launched into Outer Space (Registration Convention) enters into force on September 15.

1978 - January 24: Kosmos 954 reenters uncontrolled over northern Canada, scattering radioactive debris. - The Kessler-Cour-Palais paper is published in the Journal of Geophysical Research, identifying the cascade dynamic.

1979 - July: Skylab reenters uncontrolled, with debris falling in Western Australia.

1985 - September 13: U.S. ASM-135 missile destroys the Solwind P78-1 satellite at 345 miles (555 km), producing 285 cataloged fragments.

1993 - The Inter-Agency Space Debris Coordination Committee (IADC) is established.

1996 - July 24: Cerise satellite is struck by a fragment of an Ariane 1 upper stage, the first confirmed accidental hypervelocity collision.

2001 - March: Mir space station undergoes controlled deorbit into the South Pacific Ocean.

2002 - The original IADC Space Debris Mitigation Guidelines are adopted.

2007 - January 11: China conducts a direct-ascent anti-satellite test against Fengyun-1C at 537 miles (865 km), producing 3,500+ cataloged fragments. - February 19: Russian Briz-M upper stage explosion adds approximately 1,000 cataloged fragments to LEO. - The UN Committee on the Peaceful Uses of Outer Space (COPUOS) endorses the IADC mitigation guidelines.

2008 - February 20: U.S. Operation Burnt Frost destroys USA-193 at 153 miles (247 km), producing approximately 175 fragments that decay rapidly.

2009 - February 10: Iridium 33 collides with Cosmos 2251 at 490 miles (789 km), the first known accidental hypervelocity collision between intact satellites, producing approximately 1,800 cataloged fragments. - The U.S. Combined Space Operations Center is established at Vandenberg.

2015 - November 25: NOAA-16 fragmentation produces 275 cataloged fragments at 528 miles (850 km).

2016 - March 26: JAXA's Hitomi satellite breakup, the first known fragmentation caused by attitude control software failure.

2018 - April 2: Tiangong-1 reenters uncontrolled over the South Pacific Ocean.

2019 - March 27: India conducts Mission Shakti anti-satellite test against Microsat-R at 175 miles (282 km), producing approximately 130 fragments that decay rapidly. - April 7: INTELSAT 29e fragmentation at geostationary altitude. - The COPUOS Long-term Sustainability Guidelines are adopted.

2020 - February: MEV-1 docks with Intelsat 901, the first commercial in-orbit servicing operation. - March: Space Fence at Kwajalein achieves operational capability. - May 11: First Long March 5B core stage uncontrolled reentry, with fragments recovered in Côte d'Ivoire. - October 13: Artemis Accords opened for signature.

2021 - February 19: ClearSpace-1 prime contract awarded by ESA. - March: Astroscale ELSA-d mission launched. - April: MEV-2 docks with Intelsat 10-02. - May 9: Second Long March 5B core stage uncontrolled reentry. - November 15: Russia conducts Cosmos 1408 anti-satellite test, producing 1,500+ fragments at 298 miles (480 km). The ISS crew takes shelter in the return spacecraft.

2022 - April: U.S. Vice President announces unilateral commitment to refrain from destructive direct-ascent anti-satellite testing. - July 30: Third Long March 5B core stage uncontrolled reentry, fragments recovered in Borneo. - September: U.S. FCC adopts five-year disposal rule (FCC 22-74). - November 4: Fourth Long March 5B core stage uncontrolled reentry over the Pacific. - November 12: Long March 6A upper stage breakup at 435-497 miles (700-800 km), producing 350 cataloged fragments. - The IADC Space Debris Mitigation Guidelines have been revised.

2023 - Murphy et al. publish a PNAS paper documenting spacecraft-derived metals in stratospheric aerosols. - World Radio Communication Conference adopts decisions on large constellations.

2024 - February 26: Astroscale ADRAS-J launched. - March: ISS battery pallet reentry produces a fragment recovered in a Florida home. - April: ClearSpace-1 retargeted from Vespa to PROBA-1. - May: SpaceX Crew Dragon trunk fragments recovered in North Carolina. - June 26: Resurs-P1 breakup; fragments pass near ISS, requiring crew shelter. - July: NOAA Office of Space Commerce assumes lead role for civilian conjunction services.

2025 - Vera C. Rubin Observatory begins full Legacy Survey of Space and Time operations. - NorthStar Earth and Space deploys the first three operational space-based optical SSA satellites.

2026 - January: Astroscale announces PRELUDE mission for 2027 launch. - April: Amazon Leo (renamed from Project Kuiper in November 2025) reaches 302 production satellites in orbit. - ESA Space Environment Report documents approximately 40,000 tracked objects, 11,000 active payloads.

Appendix B. Major International Space Treaties

Treaty on Principles Governing the Activities of States in the Exploration and Use of Outer Space, including the Moon and Other Celestial Bodies (Outer Space Treaty) - Opened for signature: January 27, 1967 - Entered into force: October 10, 1967 - Parties: 115 states (mid-2026) - Principal provisions: prohibition of national appropriation of celestial bodies; prohibition of weapons of mass destruction in orbit; state responsibility for national space activities; state liability for damage caused by space objects; freedom of scientific investigation; assistance to astronauts in distress

Agreement on the Rescue of Astronauts, the Return of Astronauts, and the Return of Objects Launched into Outer Space (Rescue Agreement) - Opened for signature: April 22, 1968 - Entered into force: December 3, 1968 - Parties: 99 states (mid-2026) - Principal provisions: obligation to assist astronauts in distress; obligation to return astronauts and space objects to launching state

Convention on International Liability for Damage Caused by Space Objects (Liability Convention) - Opened for signature: March 29, 1972 - Entered into force: September 1, 1972 - Parties: 98 states (mid-2026) - Principal provisions: absolute liability for surface damage and damage to aircraft in flight; fault-based liability for damage in space; Claims Commission procedures for resolution of disputes

Convention on Registration of Objects Launched into Outer Space (Registration Convention) - Opened for signature: January 14, 1975 - Entered into force: September 15, 1976 - Parties: 75 states (mid-2026) - Principal provisions: requirement to maintain national registries of space objects and to provide registration information to the United Nations Office for Outer Space Affairs

Agreement Governing the Activities of States on the Moon and Other Celestial Bodies (Moon Agreement) - Opened for signature: December 18, 1979 - Entered into force: July 11, 1984 - Parties: 18 states (mid-2026) - Principal provisions: common heritage of mankind principle for celestial bodies; prohibition of placing weapons in lunar orbit; environmental protection requirements; international regime for resource exploitation - Note: No major spacefaring state has ratified this agreement

Artemis Accords (non-binding statement of principles) - Opened for signature: October 13, 2020 - Signatories: 50+ states (mid-2026) - Principal provisions: peaceful purposes; transparency; interoperability; emergency assistance; registration; release of scientific data; preservation of historic sites; space resources; deconfliction; orbital debris

Appendix C. Acronyms and Abbreviations

ADR — Active Debris Removal

ADRAS-J — Active Debris Removal by Astroscale-Japan (mission)

AI — Artificial Intelligence

ALMA — Atacama Large Millimeter Array

ASAT — Anti-satellite (weapon or test)

CDM — Conjunction Data Message

CNES — Centre National d'Études Spatiales (French space agency)

CNSA — China National Space Administration

COPUOS — Committee on the Peaceful Uses of Outer Space (United Nations)

CSpOC — Combined Space Operations Center

DARPA — Defense Advanced Research Projects Agency (U.S.)

DOD — Department of Defense (U.S.)

ELSA-d — End-of-Life Services by Astroscale-demonstration (mission)

ELSA-M — End-of-Life Services by Astroscale-Multi-client (mission)

EOL — End of Life

EPFL — École Polytechnique Fédérale de Lausanne

ESA — European Space Agency

EU — European Union

EUSST — European Union Space Surveillance and Tracking

EVA — Extravehicular Activity

FAA — Federal Aviation Administration (U.S.)

FCC — Federal Communications Commission (U.S.)

GEO — Geostationary Earth Orbit

GTO — Geostationary Transfer Orbit

IADC — Inter-Agency Space Debris Coordination Committee

ISO — International Organization for Standardization

ISRO — Indian Space Research Organization

ISS — International Space Station

ITU — International Telecommunication Union

JAXA — Japan Aerospace Exploration Agency

LEO — Low Earth Orbit

LTS — Long-term Sustainability (Guidelines)

MEO — Medium Earth Orbit

MEV — Mission Extension Vehicle

NASA — National Aeronautics and Space Administration (U.S.)

NOAA — National Oceanic and Atmospheric Administration (U.S.)

NRHO — Near Rectilinear Halo Orbit

OST — Outer Space Treaty

PALMS — Particle Analysis by Laser Mass Spectrometry

PMD — Post-Mission Disposal

PNAS — Proceedings of the National Academy of Sciences

PNT — Position, Navigation, and Timing

PoC — Probability of Collision

PRELUDE — Astroscale's commercial debris removal mission (2027)

RPO — Rendezvous and Proximity Operations

SDA — Space Domain Awareness

SKA — Square Kilometer Array

SPOUA — South Pacific Ocean Uninhabited Area (Point Nemo region)

SSA — Space Situational Awareness

SSO — Sun-Synchronous Orbit

SSR — Space Sustainability Rating

STM — Space Traffic Management

TLE — Two-Line Element (orbital element format)

UNCOPUOS — United Nations Committee on the Peaceful Uses of Outer Space

UNDRIP — United Nations Declaration on the Rights of Indigenous Peoples

UNOOSA — United Nations Office for Outer Space Affairs

USDV — U.S. Deorbit Vehicle (for ISS deorbit)

USSF — United States Space Force

VLA — Very Large Array

WB-57 — NASA high-altitude research aircraft

Glossary of Terms

Active Debris Removal (ADR). The deliberate capture and disposal of derelict satellites or large debris fragments by purpose-built spacecraft is intended to lower the existing debris population in supercritical altitude bands.

Apogee. The point in an elliptical orbit farthest from the central body. For Earth orbits, the apogee altitude determines the maximum altitude reached by the object during its orbital motion.

Argument of Perigee. The orbital element specifying the angular position of the perigee within the orbital plane, measured from the ascending node.

Artemis Accords. A non-binding statement of principles for cooperative lunar exploration, opened for signature in October 2020, addresses peaceful purposes, transparency, interoperability, debris mitigation, and other dimensions of space cooperation.

Atmospheric Drag. The aerodynamic force exerted on satellites by the residual atmosphere at their orbital altitude. Drag causes orbital decay over time, particularly for satellites in low Earth orbit below approximately 600 kilometers (373 miles).

Ballistic Coefficient. A measure of an object's resistance to atmospheric drag, defined as the ratio of mass to the product of cross-sectional area and drag coefficient. A higher ballistic coefficient implies less drag effect per unit mass.

Cascade. The hypothetical exponential growth of debris through a sequence of collisions, where the fragments from each collision become the targets of subsequent collisions. The Kessler Syndrome describes this dynamic.

Catastrophic Collision. In the technical literature on debris, a collision is described in which the impact energy per unit target mass exceeds approximately 40 J/g, producing complete fragmentation of both objects.

Cislunar Space. The region of space between Earth and the Moon, including various Earth-Moon Lagrange points, lunar orbits, and trajectories connecting Earth-orbit and lunar regions.

Conjunction. A predicted close approach between two cataloged objects in orbit. Conjunctions are screened systematically through tracking and analytical processes, with warnings issued when predicted close approaches meet specific criteria.

Conjunction Data Message (CDM). A standardized format for communicating conjunction information from screening services to satellite operators. Each CDM includes the predicted positions and velocities of the two objects at closest approach, the position uncertainties, and the computed probability of collision.

Controlled Reentry. A planned reentry of a spacecraft directed by deliberate propulsive maneuvers, typically targeting an unpopulated area such as the South Pacific Ocean Uninhabited Area.

Cross-section. The effective area of an object as seen from a particular direction, used in calculating collision probabilities and atmospheric drag effects.

Dark Sky. A sky condition with minimal light pollution that allows observation of natural celestial features, including stars and galaxies.

Debris. Any non-functional object in orbit, including spent rocket stages, defunct satellites, fragments from collisions or breakups, and mission-related debris.

Delta-V. The change in velocity required for a specific maneuver. Delta-V budgets are calculated for missions to determine propellant requirements.

Design for Demise. A spacecraft design approach that maximizes the fraction of mass disintegrated during reentry, intended to reduce casualty risk from surviving fragments.

Dilution. The phenomenon in which the conjunction probability decreases as position uncertainty increases, even though the actual collision risk does not necessarily decrease. Operators must account for dilution effects to avoid being misled by artificially low probability values.

Direct-Ascent Anti-satellite Weapon. An anti-satellite weapon launched on a single ascent trajectory from the ground, surface, or aircraft, targeting a satellite at a specific point in its orbit. The 1985 ASM-135, the 2007 SC-19, the 2008 SM-3, the 2019 Mission Shakti, and the 2021 Nudol systems are examples.

Earth-Moon Lagrange Points. Five specific points in the cislunar regime where the gravitational influences of Earth and the Moon balance, allowing satellites to maintain stable or quasi-stable positions with limited propulsive expenditure.

Eccentricity. The orbital element specifying the deviation of an elliptical orbit from a circular shape. Eccentricity ranges from 0 (circular) to values approaching 1 (extremely elongated).

End of Life (EOL). The conclusion of a satellite's operational mission, after which the satellite must either be disposed of through reentry, transferred to a graveyard orbit, or otherwise managed to limit its contribution to the debris environment.

Fragmentation Event. An event that produces multiple fragments from a previously intact object, including collisions, deliberate breakups, and accidental explosions due to propellant or battery failures.

Geostationary Orbit (GEO). A circular orbit in the equatorial plane at approximately 22,236 miles (35,786 kilometers) altitude, where the orbital period matches Earth's rotation period, and the satellite appears stationary relative to the surface.

Geostationary Transfer Orbit (GTO). An elliptical orbit is used to transfer payloads from low Earth orbit altitudes to geostationary altitudes. The perigee is typically in low Earth orbit, with the apogee at geostationary altitude.

Graveyard Orbit. An orbital regime designated for the disposal of decommissioned satellites, particularly used for geostationary satellites that cannot be deorbited within reasonable propellant budgets.

Hypervelocity Impact. A collision occurring at velocities sufficient to produce shock effects in target materials, typically above 0.62-1.9 miles per second (1-3 kilometers per second). Most orbital collisions occur at hypervelocity.

In-Orbit Servicing. The deliberate inspection, refueling, repair, or modification of operational satellites by other spacecraft.

Inclination. The orbital element specifying the angle between the orbital plane and Earth's equatorial plane.

Kessler Syndrome. The cascade dynamic described by Donald Kessler and Burton Cour-Palais in 1978, in which a sufficiently dense debris population can produce exponentially growing fragment populations through internal collisions.

Launching State. Under the Outer Space Treaty and Liability Convention, the state from whose territory or facility a space object is launched, or the state that procures the launch, is the responsible party. The launching state keeps responsibility and liability for the space object.

Liability Convention. The 1972 Convention on International Liability for Damage Caused by Space Objects establishes absolute liability for surface damage and fault-based liability for damage in space.

Long-term Sustainability (LTS) Guidelines. The 2019 COPUOS guidelines on the long-term sustainability of outer space activities address regulatory frameworks, safety of space operations, international cooperation, and capacity building.

Low Earth Orbit (LEO). The orbital regime is between approximately 200 kilometers (124 miles) altitude (the lower limit imposed by atmospheric drag) and approximately 2,000 kilometers (1,243 miles) altitude.

Medium Earth Orbit (MEO). The orbital regime is between approximately 2,000 kilometers (1,243 miles) and approximately 35,786 kilometers (22,236 miles) altitude (geostationary).

Mitigation. The prevention of new debris generation through systematic spacecraft design and operational practices, including end-of-life disposal, passivation, conjunction avoidance, and prevention of operational debris release.

Near Rectilinear Halo Orbit (NRHO). A specific class of cislunar orbit selected for the Lunar Gateway, characterized by a long, looping trajectory around the Earth-Moon L2 point.

Orbital Decay. The gradual reduction of orbital altitude due to atmospheric drag eventually results in atmospheric reentry.

Orbital Elements. The set of parameters specifying an orbit, including semi-major axis (or orbital period), eccentricity, inclination, right ascension of ascending node, argument of perigee, and mean anomaly.

Orbital Regime. A region of orbital space characterized by specific altitude, inclination, and dynamic properties. Major regimes include LEO, MEO, GEO, GTO, and the cislunar regimes.

Outer Space Treaty (OST). The 1967 Treaty on Principles Governing the Activities of States in the Exploration and Use of Outer Space is the foundational instrument of space law.

Passivation. The systematic removal of stored energy from spacecraft and rocket stages at the end of the mission eliminates the principal sources of post-mission explosions. Includes propellant venting, battery discharge, and gas pressure release.

Perigee. The point in an elliptical orbit closest to the central body. For Earth orbits, the perigee altitude determines the minimum altitude reached by the object during its orbital motion.

Probability of Collision (PoC). A statistical metric expressing the likelihood that two objects will collide during a predicted close approach. The PoC depends on the predicted miss distance, the position uncertainties, and the combined cross-sectional area of the objects.

Propagation. The numerical computation of an object's future position based on its current orbital elements and applicable force models.

Radar Cross-Section. The effective area of an object as observed by a radar system. Larger objects, or those with reflective surfaces, have larger radar cross-sections.

Reentry. The descent of an orbital object through the atmosphere, with associated aerodynamic heating that typically destroys most of the object's mass.

Registration Convention. The 1976 Convention on Registration of Objects Launched into Outer Space requires launching states to maintain national registries and provide registration information to the United Nations.

Right Ascension of Ascending Node. The orbital element specifying the angular position of the orbit's ascending node (where the orbit crosses Earth's equatorial plane from south to north) is measured from a fixed reference direction.

Rendezvous and Proximity Operations (RPO). The maneuvers required to bring two spacecraft to close range include final approach, station-keeping, and capture or docking.

Semi-major Axis. The orbital element specifying the half-length of the longest axis of an elliptical orbit. The semi-major axis determines the orbital period through Kepler's third law.

Space Object. Under the Outer Space Treaty, a satellite, spacecraft, rocket stage, or other artificial object launched into outer space, including any debris derived from such an object.

Space Situational Awareness (SSA). The systematic monitoring of objects and activities in space, including tracking of satellites and debris, characterization of operational status, and prediction of conjunctions and other events.

Space Surveillance Network. The U.S. global network of radars and optical sensors that maintains the principal U.S. catalog of orbital objects is operated by the Eighteenth Space Defense Squadron at Vandenberg Space Force Base.

Space Traffic Management (STM). The comprehensive coordination of orbital activity to ensure safety and sustainability, including elements of tracking, conjunction screening, regulatory oversight, and operational coordination.

Spatial Density. The number of objects per unit volume in a specified orbital regime.

Stratospheric Aerosols. Small particles suspended in the stratosphere, including natural sulfate aerosols, volcanic emissions, and increasingly, metallic particles from spacecraft reentries.

Sun-Synchronous Orbit (SSO). An orbit with an inclination near 98 degrees, designed so that the satellite's orbital plane maintains a fixed orientation relative to the sun, providing consistent lighting conditions for Earth observation.

Two-Line Element (TLE). A standard format for orbital elements, originally developed for U.S. military tracking and now used as the principal exchange format for orbital information.

Whipple Shield. A multi-layer protective structure, originally proposed by Fred Whipple in 1947, is designed to disperse the impact energy of small fragments through interaction with a thin outer plate. Stuffed Whipple shields and multi-shock variants provide enhanced protection.

Index

About the Author

Jason P. Crutchfield has spent his life paying attention to the systems most people never see. As a child, he watched the night sky and asked the questions that would eventually shape his career. How does it all stay up there? What happens when it comes down? Who is responsible for the things we leave behind? Those questions never went away. They followed him from his earliest days in the United States Army, where he served with the 2nd Battalion, 75th Ranger Regiment, through a career spanning more than three decades of senior leadership across the United States Intelligence Community, the Department of Defense, and the Department of Energy. That work taught him how complex systems succeed and fail, and how the most important risks are often the ones nobody is talking about yet.

The orbital debris problem fits that pattern exactly. By the time Jason began writing this book, the data was unmistakable. Tens of thousands of tracked objects, millions of untracked fragments, and a launch rate accelerating faster than any mitigation strategy could absorb. Yet the conversation about it remained scattered across academic papers, classified briefings, and policy documents that the people most affected, including engineers, operators, regulators, taxpayers, and citizens, could not easily access. Jason wrote Space Debris: Clearing the Final Frontier to close that gap. Drawing on dozens of sources and current data through 2026, the book examines the technological, economic, and policy dimensions of one of the most urgent challenges facing the modern space environment.

Outside the office, Jason has spent many years underwater. He has worked as a commercial diver in the Gulf of Mexico and Alaska and has taught as a certified scuba diving instructor throughout his long career on the water. That experience has shaped how he thinks and writes about pressure, the environment, and the discipline required to operate safely in places that punish carelessness. Whether the environment is 1,000 feet underwater or 1,200 miles up, the same principles apply. Respect physics. Plan the mission. Leave the place better than you found it. His forthcoming companion volume, Economies of Space Debris, extends the analysis in this book into the financial frameworks now emerging around active debris removal and orbital sustainability. He is committed to developing a debris policy as an essential component of broader space sustainability. He welcomes engagement with readers, fellow researchers, and practitioners on the topics addressed in this volume.

Jason holds an MBA from Duke University and a Bachelor of Science from the University of Alaska. This volume reflects his sustained effort to integrate the diverse technical, legal, regulatory, economic, and operational dimensions of the orbital debris challenge into a coherent framework accessible to general readers, practitioners, and academic researchers. The work draws on extensive engagement with the published technical and policy literature, the operational practices of major space programs, and the developing international framework for space sustainability.

Editorial Note

This edition of "Space Debris: Clearing the Final Frontier" reflects extensive editorial development from the original manuscript. The objective has been to produce a comprehensive reference that meets professional publishing standards for both academic and general readership. Key editorial activities have included:

Comprehensive fact-checking against current authoritative sources, including the European Space Agency's 2025 Space Environment Report, recent peer-reviewed publications, regulatory documents, and primary mission reports. Statistics and operational details have been verified against the most recent available data as of mid-2026, with the understanding that the orbital debris environment continues to evolve rapidly.

Standardization of voice and style throughout the manuscript. The text employs a measured technical register suitable for both academic and general readership, with consistent application of terminology, formatting conventions, and citation style.

Integration of recent developments, including the November 2025 renaming of Project Kuiper to Amazon Leo, the April 2024 retargeting of ClearSpace-1 from Vespa to PROBA-1, the January 2026 announcement of the PRELUDE mission, the 2024 Crew Dragon trunk reentry over North Carolina, the 2024 ISS battery pallet reentry incident, the September 2022 FCC five-year disposal rule, and other recent developments that shape current understanding of the field.

Comprehensive treatment of the cumulative environmental dimensions of space activity, including the 2023 NOAA stratospheric aerosol findings, the constellation impact on ground-based astronomy, and the developing concept of orbital space as a common. These dimensions, less prominent in earlier treatments of the debris problem, have emerged as important considerations in contemporary policy and practice.

Detailed examination of the cascade analysis as the analytical foundation of contemporary debris policy. The Kessler-Cour-Palais 1978 paper and its subsequent development through more sophisticated modeling provide the technical basis for the broader response, and the manuscript treats this foundation in sufficient depth to support readers' independent assessment of the technical claims.

Inclusion of comprehensive front matter and back matter to support the reference function of the work. Front matter provides context and orientation; back matter provides the apparatus for navigation, verification, and further investigation.

The orbital debris challenge continues to develop, and any reference work in this field can capture only a snapshot of a dynamic situation. The reader is encouraged to consult current sources for the latest data, particularly the ESA Space Environment Report (updated annually), the IADC working group reports (updated as the committee's work proceeds), the COPUOS documentation (available through the United Nations Office for Outer Space Affairs), and the various national regulatory documents that govern current practice.

Comments, corrections, and suggestions for future editions are welcome. The field's development will continue, and the response to the challenge depends on informed readers' engagement with both the published literature and ongoing institutional and policy work.

References

The following references support the factual claims, technical content, and policy analysis throughout this volume. References are formatted in APA 7th edition style. Where multiple editions or annual updates exist, the most recent edition consulted is cited; readers seeking historical data should consult earlier editions of the relevant reports.

Adamopoulos, J., & Madry, S. (2023). Commercial space situational awareness: Market analysis and forecast. Northern Sky Research Reports.

Adushkin, V. V., Aksenov, O. Y., Veniaminov, S. S., Kozlov, S. I., & Tyurenkova, V. V. (2020). The small orbital debris population and its impact on space activities and ecological safety. Acta Astronautica, 176, 591–597. https://doi.org/10.1016/j.actaastro.2020.01.015

Anselmo, L., & Pardini, C. (2008). Space debris mitigation in geosynchronous orbit. Advances in Space Research, 41(7), 1091–1099. https://doi.org/10.1016/j.asr.2007.04.082

Anselmo, L., & Pardini, C. (2024). Long-term evolution of the orbital debris environment in low Earth orbit under contemporary mitigation requirements. Acta Astronautica, 215, 358–370.

Astroscale Holdings. (2024). ADRAS-J mission summary report: Phase 1 rendezvous and inspection of H-IIA upper stage. Astroscale Holdings, Inc. https://astroscale.com

Astroscale Holdings. (2026, January 15). Astroscale announces PRELUDE commercial debris removal mission for 2027 launch [Press release]. https://astroscale.com/press

Bastida Virgili, B., Dolado, J. C., Lewis, H. G., Radtke, J., Krag, H., Revelin, B., Cazaux, C., Colombo, C., Crowther, R., & Metz, M. (2016). Risk to space sustainability from large constellations of satellites. Acta Astronautica, 126, 154–162. https://doi.org/10.1016/j.actaastro.2016.03.034

Boley, A. C., & Byers, M. (2021). Satellite mega-constellations pose risks in low Earth orbit, in the atmosphere, and on Earth. Scientific Reports, 11, 10642. https://doi.org/10.1038/s41598-021-89909-7

Burch, J. (2021). The legal and ethical implications of orbital debris removal. Journal of Space Law, 44(2), 243–278.

ClearSpace SA. (2024). ClearSpace-1 mission status update following PROBA-1 retargeting. ClearSpace SA. https://clearspace.today

CONFERS. (2023). Recommended design and operational practices for on-orbit servicing—Consortium for Execution of Rendezvous and Servicing Operations.

Crowther, R. (2002). Space junk: Protecting space for future generations. Science, 296(5571), 1241–1242. https://doi.org/10.1126/science.1071335

Defense Advanced Research Projects Agency. (2022). Hallmark program final report. DARPA Space Domain Awareness Office.

European Commission. (2021). EU approach to space traffic management. Joint Communication JOIN(2022) 4 final.

European Space Agency. (2023). Space debris mitigation policy for ESA projects (ESA/ADMIN/IPOL(2014)2). European Space Agency.

European Space Agency. (2025). ESA's annual space environment report 2025. ESA Space Debris Office. https://www.esa.int/Space_Safety/Space_Debris

European Union. (2021). Regulation (EU) 2021/696 of the European Parliament and of the Council of 28 April 2021 establishing the Union Space Programme. Official Journal of the European Union, L 170, 69–148.

Federal Aviation Administration. (2024). Commercial space transportation regulations and policies. FAA Office of Commercial Space Transportation. https://www.faa.gov/space

Federal Communications Commission. (2022). Mitigation of orbital debris in the new space age (Report and Order, FCC 22-74). Federal Communications Commission.

Federal Communications Commission. (2024). Space innovation: Streamlining licensing procedures for satellite applications (FCC 24-32). Federal Communications Commission.

Foust, J. (2024, May 14). Crew Dragon trunk fragments recovered in North Carolina. SpaceNews. https://spacenews.com

Geppert, U., Diekmann, B., & Daul, T. (2014). On the lifetime of debris generated in the geostationary belt. Advances in Space Research, 53(8), 1183–1191.

Gleghorn, G., Asay, J., Atkinson, D., Flury, W., Johnson, N., Kessler, D., Knowles, S., Rex, D., Toda, S., Veniaminov, S., & Warren, R. (1995). Orbital debris: A technical assessment. National Academy Press.

Gorman, A. (2014). The anthropocene in the solar system. Journal of Contemporary Archaeology, 1(1), 87–91.

Hakes, A., & Lemmens, S. (2020). The classification of breakup events for the European Space Agency space debris program. Journal of Space Safety Engineering, 7(3), 364–369.

Hall, R. C. (2014). The history of NASA's exploration of low Earth orbit. NASA History Division.

Henry, C. (2020, July 22). Northrop Grumman's MEV-1 servicer demonstrates commercial in-orbit servicing in GEO. SpaceNews.

Inter-Agency Space Debris Coordination Committee. (2021). IADC space debris mitigation guidelines (IADC-02-01, Revision 3). https://iadc-home.org

International Astronomical Union. (2022). The IAU Centre for the Protection of the Dark and Quiet Sky from Satellite Constellation Interference: Founding charter and operational plan. International Astronomical Union.

International Organization for Standardization. (2019). Space systems, Space debris mitigation requirements (ISO 24113:2019)—International Organization for Standardization.

International Telecommunication Union. (2023). Final acts of the World Radiocommunication Conference (WRC-23). International Telecommunication Union.

Jah, M. K. (2018). Space situational awareness for sustainable space operations. The Bridge: Journal of the National Academy of Engineering, 48(3), 14–22.

Japan Aerospace Exploration Agency. (2024). Space activities act compliance manual for commercial operators (Rev. 4). JAXA Space Tracking and Communications Center.

Johnson, N. L. (2010). Orbital debris: The growing threat to space operations. Advances in the Astronautical Sciences, 137, 3–11.

Johnson, N. L., Krisko, P. H., Liou, J.-C., & Anz-Meador, P. D. (2001). NASA's new breakup model of EVOLVE 4.0. Advances in Space Research, 28(9), 1377–1384.

Kessler, D. J., & Cour-Palais, B. G. (1978). Collision frequency of artificial satellites: The creation of a debris belt. Journal of Geophysical Research: Space Physics, 83(A6), 2637–2646. https://doi.org/10.1029/JA083iA06p02637

Kessler, D. J., Johnson, N. L., Liou, J.-C., & Matney, M. (2010). The Kessler syndrome: Implications for future space operations. Advances in the Astronautical Sciences, 137, 47–61.

Klinkrad, H. (2006). Space debris: Models and risk analysis. Springer-Verlag. https://doi.org/10.1007/3-540-37674-7

Klinkrad, H., & Johnson, N. L. (2009). Space debris environment remediation concepts. Acta Astronautica, 65(11–12), 1655–1668.

Krag, H., Lemmens, S., & Letizia, F. (2017). Space traffic management through the control of the space environment's capacity—proceedings of the 1st IAA Conference on Space Situational Awareness.

Kucharski, D., Kirchner, G., Lim, H.-C., Otsubo, T., Bennett, J. C., Lachut, M., Hwang, J. Y., Kim, S., & Park, M. (2017). Photometric studies of rapidly spinning decommissioned GEO satellite METEOSAT-7. Acta Astronautica, 139, 233–238.

Liou, J.-C. (2011). An active debris removal parametric study for LEO environment remediation. Advances in Space Research, 47(11), 1865–1876.

Liou, J.-C., Anilkumar, A. K., Bastida Virgili, B., Hanada, T., Krag, H., Lewis, H., Raj, M. X. J., Rao, M. M., Rossi, A., & Sharma, R. K. (2018). Stability of the future LEO environment: An IADC comparison study. Acta Astronautica, 138, 376–385. https://doi.org/10.1016/j.actaastro.2017.06.027

Liou, J.-C., & Johnson, N. L. (2006). Risks in space from orbiting debris. Science, 311(5759), 340–341. https://doi.org/10.1126/science.1121337

Liou, J.-C., & Johnson, N. L. (2008). Instability of the present LEO satellite populations. Advances in Space Research, 41(7), 1046–1053.

Lyall, F., & Larsen, P. B. (2018). Space law: A treatise (2nd ed.). Routledge.

Macdonald, M., & Badescu, V. (Eds.). (2014). The international handbook of space technology. Springer.

McDowell, J. (2020). The low Earth orbit satellite population and impacts of the SpaceX Starlink constellation: the Astrophysical Journal Letters, 892(2), L36.

McKnight, D., Maher, R., & Nagl, L. (1995). Refined algorithms for structural breakup due to hypervelocity impact. International Journal of Impact Engineering, 17(4–6), 547–558.

McKnight, D., Witner, R., Letizia, F., Lemmens, S., Anselmo, L., Pardini, C., Rossi, A., Kunstadter, C., Kawamoto, S., Aslanov, V., Dolado, J. C., Schaub, H., & Slatton, Z. (2021). Identifying the 50 statistically most concerning derelict objects in LEO. Acta Astronautica, 181, 282–291.

Megna, B. (2024). Commercial space station development: A status review of the post-ISS commercial LEO destinations program. NASA Office of Inspector General.

Murphy, D. M., Abou-Ghanem, M., Cziczo, D. J., Froyd, K. D., Jacquot, J., Lawler, M. J., Maloney, C., Plane, J. M. C., Ross, M. N., Schill, G. P., & Shen, X. (2023). Metals from spacecraft reentry in stratospheric aerosol particles. Proceedings of the National Academy of Sciences, 120(43), e2313374120. https://doi.org/10.1073/pnas.2313374120

National Aeronautics and Space Administration. (2024a). NASA awards SpaceX $843 million contract for U.S. Deorbit Vehicle development [Press release]. NASA. https://www.nasa.gov

National Aeronautics and Space Administration. (2024b). NASA orbital debris program office: Quarterly news, 28(1–4). NASA Johnson Space Center.

National Aeronautics and Space Administration. (2024c). NASA technical standard 8719.14: Process for limiting orbital debris (Rev. C). NASA.

National Oceanic and Atmospheric Administration. (2024). Civilian space situational awareness services transition plan. U.S. Department of Commerce, Office of Space Commerce.

National Research Council. (2011). Limiting future collision risk to spacecraft: An assessment of NASA's meteoroid and orbital debris programs. National Academies Press. https://doi.org/10.17226/13244

NASA Office of Inspector General. (2021). NASA's efforts to mitigate the risks posed by orbital debris (Report No. IG-21-011). NASA.

NorthStar Earth and Space. (2025). Constellation deployment update: First operational satellites achieve milestones [Press release]. NorthStar Earth and Space.

Office of Science and Technology Policy. (2022). National orbital debris implementation plan. Executive Office of the President.

Pardini, C., & Anselmo, L. (2018). Evaluating the impact of space activities in low Earth orbit. Acta Astronautica, 184, 11–22.

Pelton, J. N. (2015). New solutions for the space debris problem. Springer International Publishing. https://doi.org/10.1007/978-3-319-17151-7

Pelton, J. N., & Madry, S. (Eds.). (2020). Handbook of small satellites: Technology, design, manufacture, applications, economics, and regulation. Springer International Publishing.

Roscosmos State Corporation. (2024). Russian space surveillance system: Operational status report. Roscosmos.

Ross, M., Sheaffer, P. (2014). Radiative forcing caused by rocket engine emissions. Earth's Future, 2(4), 177–196.

Schaub, H., Jasper, L. E. Z., Anderson, P. V., & McKnight, D. S. (2015). Cost and risk assessment for spacecraft operational decisions due to the space debris environment. Acta Astronautica, 113, 66–79.

Schildknecht, T. (2007). Optical surveys for space debris. The Astronomy and Astrophysics Review, 14(1), 41–111.

Secure World Foundation. (2024). Global counterspace capabilities: An open source assessment. Secure World Foundation.

Shoemaker, M. A., Vavrin, A. B., Whittle, A. C., & Wright, M. M. (2023). Statistical evaluation of break-up events as compiled in the NASA Standard Satellite Breakup Model. Proceedings of the 8th European Conference on Space Debris.

Society of Satellite Professionals International. (2024). State of the Satellite Industry Report 2024—Bryce Tech for SSPI.

Space Foundation. (2025). The space report 2025. Space Foundation.

SpaceX. (2024). Starlink space sustainability report. Space Exploration Technologies Corp. https://www.spacex.com

United Kingdom Space Agency. (2023). Outer Space Activities Act and the licensing of space activities. UK Space Agency.

United Nations Committee on the Peaceful Uses of Outer Space. (2019). Guidelines for the long-term sustainability of outer space activities (A/74/20, Annex II). United Nations Office for Outer Space Affairs.

United Nations Office for Outer Space Affairs. (2024). United Nations treaties and principles on outer space and related General Assembly resolutions. United Nations Office for Outer Space Affairs.

United States Space Force. (2024). Space surveillance network operations report (FY 2023). U.S. Space Force, Eighteenth Space Defense Squadron.

Weeden, B. (2010). 2009 Iridium-Cosmos collision fact sheet. Secure World Foundation.

Weeden, B., & Samson, V. (Eds.). (2024). Global counterspace capabilities: An open source assessment. Secure World Foundation.

Wiedemann, C., Krag, H., Bendisch, J., & Sdunnus, H. (2005). Analyzing costs of space debris mitigation methods. Advances in Space Research, 35(7), 1241–1248.

World Economic Forum. (2024). Space sustainability rating: Annual report 2024. World Economic Forum.

Wright, D. (2007). Space debris. Physics Today, 60(10), 35–40. https://doi.org/10.1063/1.2800252

Treaties and Legal Instruments

Convention on International Liability for Damage Caused by Space Objects, opened for signature March 29, 1972, 961 U.N.T.S. 187 (entered into force September 1, 1972) [Liability Convention].

Convention on Registration of Objects Launched into Outer Space, opened for signature January 14, 1975, 1023 U.N.T.S. 15 (entered into force September 15, 1976) [Registration Convention].

Treaty on Principles Governing the Activities of States in the Exploration and Use of Outer Space, including the Moon and Other Celestial Bodies, opened for signature January 27, 1967, 610 U.N T.S. 205 (entered into force October 10, 1967) [Outer Space Treaty].

Agreement Governing the Activities of States on the Moon and Other Celestial Bodies, opened for signature December 18, 1979, 1363 U.N.T.S. 3 (entered into force July 11, 1984) [Moon Agreement].

NASA, ESA, JAXA, CSA, ASI, UAE Space Agency, & UK Space Agency. (2020). The Artemis Accords: Principles for cooperation in the civil exploration and use of the Moon, Mars, comets, and asteroids for peaceful purposes. National Aeronautics and Space Administration. https://www.nasa.gov/specials/artemis-accords/

Note on References

Citations to news articles, press releases, and government documents reflect the most authoritative public source available at the time of writing. Where references include URLs, those URLs were verified during manuscript preparation but may change over time. Readers seeking definitive citations for legal instruments should consult the United Nations Office for Outer Space Affairs treaty database. Readers seeking the most current versions of regulatory documents should consult the issuing agency's website.

www.ingramcontent.com/pod-product-compliance
Lightning Source LLC
LaVergne TN
LVHW090350240926
848801LV00011B/164/J
* 9 7 9 8 9 9 6 1 7 5 6 0 4 *